The Australian Geographic Book of
The RED Centre

Text by Jenny Stanton

Photography by Barry Skipsey

His face daubed with traditional ochres and framed by a ceremonial head-dress, Stanley Douglas (above) prepares for one of the many ceremonies that have enriched Aboriginal life in the Red Centre for more than 30,000 years.

Title page: With Uluru in the background, Walter Pukutiwara and Tiger and Billy Wara display their ancient culture during a corroboree.

Symbols of the Centre (opposite). A lone dingo crosses the road near Kata Tjuta. Sacred to Aborigines, the distinctive domes are now visited by some 300,000 people every year.

A lone ghost gum (above) rises from the foothills of the MacDonnell Ranges. The formidable-looking MacDonnells (opposite) stretch for more than 220 km across the centre of Australia.

First published in 1995; reprinted 1998, 1999, 2000, 2001, 2002
by Australian Geographic Pty Ltd, PO Box 321,
Terrey Hills, NSW 2084, Australia. Phone: (02) 9473 6700,
Fax: (02) 9473 6701, email: books@ausgeo.com.au

Managing Director: Ken Rosebery
Managing Editor, Books: Averil Moffat

Coordinating Editor: Jenny Stanton
Design and Photographic Edit: John Witzig
Director of Cartography: Will Pringle
Production Manager: Valerie Reed
Photo Research: Cheryl Rose
Research: Tony Stanton
Editor: Frank Povah

Printed in Australia by Inprint Limited

*This page: Rains soften the rugged west MacDonnell land-
scape by carpeting the ground with a mantle of grasses.*

*Opposite: Jimmy Aitken explains the significance of
Aboriginal paintings at a "gallery" in Watarrka National Park.*

*Contents pages: Stockmen muster Brahman cattle on a
station on the northern flanks of the MacDonnell Ranges.*

National Library of Australia Cataloguing-in-Publication Data:

Stanton, Jenny
    Australian Geographic Book of The Red Centre

    Includes index.
    ISBN 1 86276 013 6

    1. Australia, Central – Guidebooks. 2. Australia, Central –
    History. I. Australian Geographic Pty Ltd. II. Title.
    III. Title: Red Centre.

919.420463

Acknowledgements

Australian Geographic thanks the Conservation Commission
of the Northern Territory (especially Stuart Traynor),
Paul Ashton of the Central Australian Tourism Industry
Association, Paul Le Messurier of the Northern Territory
Geological Survey, Sue Shearer of the Department of
Industries and Development and Bruce Strong for their
help in refereeing the book.

Jenny Stanton and Barry Skipsey thank the Northern
Territory Tourist Commission, the Territory Motor Inn,
Ayers Rock Resort, Frontier Kings Canyon, Glen Helen
Homestead, MacDonnell Range Tourist Park, Outback
Ballooning, Bob Petty of Rudolph Gunz Pty Ltd, Damien
Ryan of Alice Springs Camera Shop, Jerome Ryan of The
Foto Centre, Alice Springs, E.W. Fisher Ltd and Ken and
Jan Watts of Rural Helicopters for their support.

*Front cover: The Simpson Desert is home to 92 reptile
species, including the thorny devil (inset).*
*Back cover: Idyllic camping spots, like this one in the east
MacDonnell Ranges, are abundant in the Red Centre.*

Contents

Foreword *by Dick Smith*   11

Introduction   13

1 Landscape and Geology   15

2 Aboriginal History   25

3 European History   39

4 The Red Centre today   51

5 Wildlife   59

6 Plants and their communities   67

7 Alice Springs   75

8 West of Alice Springs   91

9 East of Alice Springs   111

10 South-West of Alice Springs   119

11 South-East of Alice Springs   143

12 North of Alice Springs   159

13 Travel Tips   171

Index   174

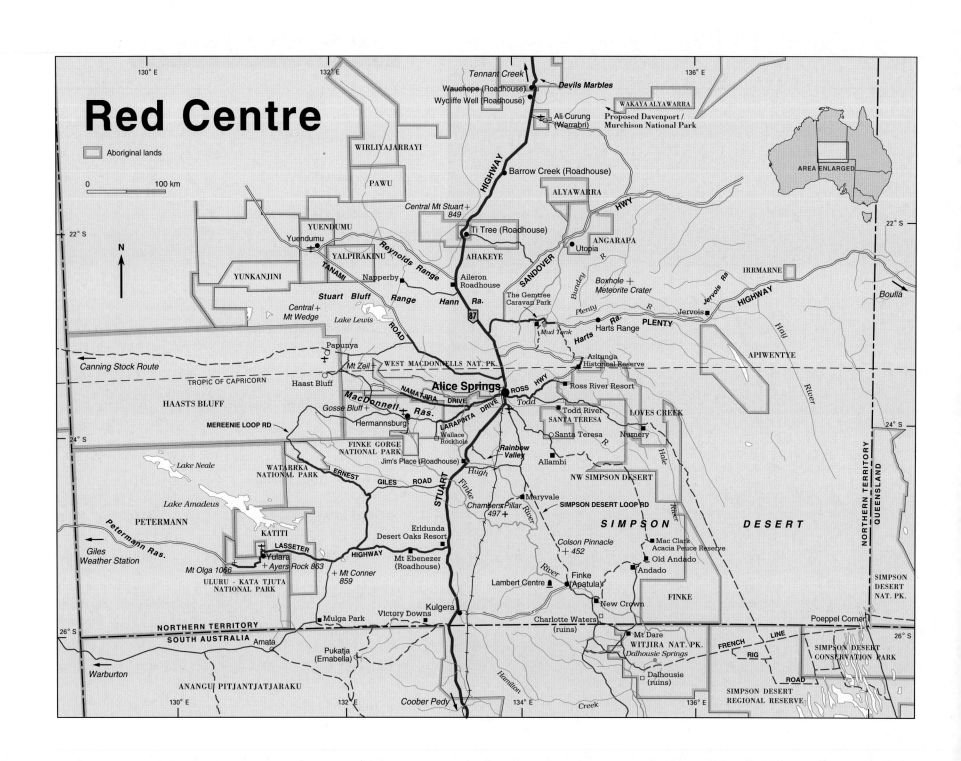

I first visited the Red Centre in July 1977 during an expedition to try to find the wreckage of the *Kookaburra*, the tiny monoplane forced down with engine trouble in the Tanami Desert on 10 April 1929.

The *Kookaburra* story is especially poignant because pilot Keith Anderson and engineer Bobby Hitchcock, who were on their way to the Kimberley to search for the *Southern Cross*, managed to fix the engine but couldn't take off because they were surrounded by thick scrub. They tried in vain to hack it down with a pocket knife but died of thirst and were later spotted by the rescued crew of the *Southern Cross*.

I didn't find the wreckage on that occasion, but I'll always remember flying south-east from the Tanami and seeing Heavitree Gap and Alice Springs, where the *Kookaburra* had taken off on its ill-fated flight. I felt more determined than ever to locate the little plane.

A year later I was successful and I'm delighted the *Kookaburra* is now on display at the Central Australian Aviation Museum in Alice Springs as a tribute to the early aviators I admire so much.

In 1985, I returned to the Red Centre on a more challenging search – for Lasseter's Reef. Harold Lasseter claimed to have discovered a rich, gold-bearing reef in 1897 but he died in 1931 while on an expedition to find it again, taking the mystery to his grave. His son Bob, who is convinced the reef exists, joined me on two thorough aerial searches of the area to the south-west of Alice Springs and although we didn't find the reef, we narrowed down the likely area and uncovered a fascinating story for the first issue of AUSTRALIAN GEOGRAPHIC.

I've been back to the Red Centre many times between those expeditions and in the years since, and I've discovered many special places and met some very special people. Perhaps my most outstanding memory is of flying my helicopter east from Alice Springs in 40°C heat in January 1990 and looking down on a raging brown torrent of water near Ross River Homestead. The flash flooding was a powerful reminder of the effect storms can have in one of the hottest, driest parts of our continent.

Other vivid memories are of climbing 1531-metre Mt Zeil, the highest point in the Territory, and later visiting the family of Eddie Connellan,

Dick Smith savours the beauty and solitude of the Simpson Desert, one of the world's great sand-ridge deserts.

PIP SMITH

who founded the Centre's first airline, at their neighbouring property, Narwietooma. I also recall travelling to Palm Valley with my friend Gus Williams of Hermannsburg and discussing the many challenges faced today by Aborigines, who have lived in the Red Centre for an incredible 30,000 years or more. And I love nothing better than listening to Ted Egan, another old friend, entertaining audiences with his outback songs and stories.

It was author Jenny Stanton, an experienced journalist and former associate editor of AUSTRALIAN GEOGRAPHIC, who suggested the Red Centre would be an ideal subject for our series of books on Australian destinations.

She spent three months exploring the region and indulging her passions for bushwalking, camping and birdwatching. She delves into the Centre's fascinating history and geography then takes you on a guided tour. Along the way you'll visit well-known attractions like Alice Springs and Uluru, and little-known spots like Ruby Gap, Boggy Hole and Chambers Pillar, where Jenny felt she was seeing the landscape much as the early European explorers did.

If Jenny's enthusiasm for the region doesn't convince you to roll up your swag, then Barry Skipsey's fantastic photography surely will. Barry loves photographing the area he now calls home and I'm sure you'll agree that his work is some of the best our books have ever featured.

I hope this impressive guide will help you to find your own special places in the Red Centre and that it will encourage you to play your part in preserving them for future generations.

Dick Smith
Chairman, Australian Geographic Society

The MacDonnell Ranges landscape that was immortalised by artist Albert Namatjira.

During my research for this book, I met a farmer in Ti Tree who told me his hobby was fishing. When I asked him where he fished, his face broke into a grin. "You're in the middle of the best spots in Australia here," he replied, then, seeing my puzzled look, he added: "To the north, the south, the east, the west."

It's undoubtedly the Red Centre's distance from the coast that sets it apart from the rest of Australia. Most of us live in cities near the sea and even Adelaide, the nearest State capital to Alice Springs, is 1337 kilometres away. For virtually everyone, a journey to our red heart is an expedition.

I first visited the Red Centre in 1991, when I took seven months' leave from AUSTRALIAN GEOGRAPHIC to explore some of Australia. I spent a month there and was captivated by what I found. Here was a rugged landscape that was also stunningly beautiful; a harsh landscape hiding a profusion of green oases; an arid landscape that contained some of the loveliest waterholes to be found anywhere. I marvelled at the resourcefulness of the Aborigines and at the tenacity of the early European explorers – and I knew that one day I'd be back.

Three years later I was, this time for two months to research this book. I discovered many new places, revisited those I'd seen before – and found most even better the second time around – and met some wonderful Territorians.

I felt there was a real need for a comprehensive guide to the region and that is what I've set out to provide. Over the next 163 pages, I'll introduce you to a 300,000 square kilometre area that stretches from the Devils Marbles in the north to the South Australian border in the south, and from Kata Tjuta in the west to Old Andado in the east. The first six chapters provide background information on the history, geography, plant and animal life of the region, the next six take you with my husband Tony and me on a tour of the main tourist areas, and the final chapter gives hints to help you plan your visit.

I'd particularly like to thank Tony for accompanying me, for doing the trickiest bits of four-wheel-driving and assisting with my research. I couldn't have undertaken this project without him. My thanks also to the people of the Red Centre who were so generous with their knowledge, their time and their hospitality, and to the many AG members I met for their enthusiasm, encouragement and friendship.

Last, but by no means least, I must say a special thank you to Barry Skipsey for his stunning photography. Barry is one of the many Territorians I met who visited the Red Centre for a holiday and decided to stay, and his love for his "backyard" shines in all he does. The professionalism and enthusiasm he brought to this project made him a joy to work with.

I hope you enjoy reading this book as much as I enjoyed researching and writing it, and that you'll be inspired to plan your own Red Centre adventure. I'm already working on my next one!

Jenny Stanton

Author Jenny Stanton and photographer Barry Skipsey swap notes in one of their many bush "offices".

Landscape and Geology
2000 million years in the making

Uluru is undoubtedly the best-known and most-visited of the Red Centre's extraordinary rock formations and visitors can learn about its origins – and those of the region's other spectacular landforms – from both an Aboriginal and a geological perspective. Aborigines believe the rock, which rises 348 m from the surrounding plain and is almost 9 km in circumference, was formed by ancestral beings who travelled the land and left details of their adventures in every nook and cranny of its

The 300,000 square kilometres of the Red Centre contain some of the most spectacular landforms on Earth – and some of the most colourful. Their brilliant oranges, reds, purples and blue-greys are all intensified by the clearest, bluest of outback skies and the strong desert sun – and appear to change colour by the hour.

The mighty MacDonnell Ranges stretch for more than 220 km and rise to 1531-metre Mt Zeil, the highest point west of the Great Dividing Range. The dunes of the Simpson Desert are up to 200 km long and cover an area more than twice the size of Tasmania.

Uluru and Kata Tjuta are no ordinary rock formations. Uluru stands 348 m higher than the surrounding plain and is nearly 9 km in circumference, and the many domes of Kata Tjuta cover an even bigger area and rise more than 600 m. What's visible is merely the tip of the iceberg: much of their bulk is buried beneath the sand.

Other remarkable features abound – from the chaotic Devils Marbles, weathered boulders up to 6 m in diameter that are scattered in precarious-looking heaps, to Tnorala, a comet crater once 2000 m high and 20 km wide, and the hundreds of mesas that stud the landscape like vast party cakes.

Amid these striking landforms are a number of paradoxes. Here is an arid landscape where the rivers rarely flow, the annual rainfall can be as low as 100 millimetres and the nearest coast is almost 1000 km away. Yet it's dotted with pockets of lush palm trees and ferns, its meagre permanent waterholes are home to 11 species of fish and are a haven for thousands of waterbirds, and its rivers can become raging torrents after just a few hours' rain.

Aborigines have found food, water and shelter within this extraordinary landscape for some 30,000 years (see Chapter 2). They believe it was formed back in the Dreamtime, or creation time, by ancestral beings who travelled the land and left details of their adventures in the form of features like the Devils Marbles, said to be the eggs of the Rainbow Serpent, and the many Dreaming trails that criss-cross the Red Centre.

Geologists have a different explanation. They believe the Earth and other planets were formed about 4600 million years ago and that some of the rocks from which the Red Centre's landforms have evolved were created nearly 2000 million years ago.

If satellite technology had been available then to record images every 50 million years, the resulting snapshots would document a landscape that *surface. They interpret these adventures through a wealth of colourful stories that have been passed on from generation to generation – and which they now share with an increasing number of visitors eager to see the landscape through their eyes. Geologists see things differently, tracing Uluru's origins back 600 million years to when the region was covered by a vast inland sea. They blame erosion by wind and rain for its many distinctive pockmarks and furrows.*

has been submerged beneath a vast inland sea, covered by ice sheets and thrown up into mountains of Himalayan proportions that have been faulted and folded, eroded by wind, rain and water, and subjected to huge variations in temperature. They'd reveal that it wasn't until just over 300 million years ago that features like Uluru and Kata Tjuta, Kings Canyon and the MacDonnell Ranges were formed, and that they only started to take on their familiar appearance 65 million years ago after more than 200 million years of erosion.

Today it's hard for visitors to picture the Red Centre submerged by water. But for most of the period between 1870 and 350 million years ago, it was covered by a shallow sea surrounded by sandy beaches. Over the millennia, layers of sand were washed into the sea and they gradually hardened to become the sandstones that are so prevalent in the area today. The Heavitree quartzite of the MacDonnell Ranges was formed in this way about 900 million years ago and is the base of a series of layers of rock that may have been an incredible 15 km thick. At Ellery Creek and near Jessie Gap, ripple marks can be seen in the rock where the sea gently lapped over it as it hardened.

The parallel ridges of the MacDonnell Ranges (opposite) reached more than 9 km – the height of the Himalaya – when they were thrown up 310 million years ago, but erosion has gradually worn them down until only their stumps remain. These are still impressive, soaring some 1000 m from the surrounding plain and rising to 1531 m Mt Zeil, the highest point west of the Great Dividing Range. Similar forces also created the breakaway country (above) south-east of Alice Springs.

HOWARD WHELAN

As the sand was gradually changed to rock, the sea became shallower and water evaporated, leaving a crust of salt. Algal mounds called stromatolites, the earliest-known form of life, grew, and layers of crystallised minerals formed, gradually hardening to become the limestone of the Bitter Springs formation, a softer rock that is readily visible in parts of the eastern MacDonnell Ranges. The fossilised remains of stromatolites can be seen near Ellery Creek Big Hole and Ross River Homestead.

Sheet ice, formed during ice ages 750 and 625 million years ago, eroded some of these rocks and re-formed them into a pebbly rock visible in parts of the west MacDonnell Ranges. But it was two major mountain-building events 600 and nearly 350 million years ago that had the most impact on this flat and largely featureless area.

The first was the Petermann Orogeny, when high, jagged mountains were pushed up in the southern margin of the sea, where the Petermann Ranges are today, as the layers of sediment were crumpled and buckled by intense north–south pressure. Subsequent weathering and erosion of these peaks produced enormous amounts of sediments that spread north and east, and eventually formed Uluru and Kata Tjuta.

Despite being formed by sediments from the same parent mountains, Uluru and Kata Tjuta are composed of very different rock. Uluru is arkose, a coarse-grained sandstone largely eroded from granite. Kata Tjuta is a conglomerate of granite and basalt boulders cemented by sand and mud – and sometimes likened to a plum pudding. Geologists say the sediments that formed Uluru are finer because they came from a very different part of the ancient range and travelled further.

Water gradually covered these sediments, which were at least 2.5 km thick, and it was not until the Alice Springs Orogeny, between 340 and 310 million years ago, that they were again exposed. The horizontal layers of Uluru arkose were tilted nearly 90 degrees as they were thrust upwards, and subsequent weathering by wind and water has highlighted this. The layers have eroded unevenly, leaving a pattern of distinctive, near-vertical ridges and furrows.

The Kata Tjuta conglomerate was tilted at an angle of about 20 degrees as it was squeezed up, suffering a number of fractures in the process. Subsequent weathering of these joints has eroded the mass into the multitude of domes that exist today.

The Alice Springs Orogeny also threw up the peaks of the MacDonnell Ranges, which reached more than 9 km – the height of the Himalaya – before they were eroded, and the ranges to the south. All run east to west because they were squeezed up by north–south pressures.

The intensity of the forces at work during this time can be seen at Ellery Creek Big Hole and Inarlanga Pass, where beds of Heavitree quartzite and Bitter Springs limestone have been buckled into tight curves, and at Glen Helen Gorge, where horizontal layers of Pacoota sandstone were tilted 90 degrees. Other legacies from this era are the many gorges carved through the ranges. Ancient rivers like the Finke, the Ellery and the Hugh pushed their way through vertical fractures, gradually widening them into the gaps and gorges that are visited by thousands of tourists every year.

Graphic evidence of the creatures that once inhabited the vast inland sea that covered much of the region was discovered by tour operator Chris "Brownie" Brown (opposite) during a 1992 AUSTRALIAN GEOGRAPHIC scientific expedition. This squid-like nautiloid had a shell more than 2 m long from which eight or 10 tentacles would have protruded. Now on display in the Australian Museum, it is more than 400 million years old – 100 million years older than 1380 m Mt Sonder (above) near the western end of the MacDonnell Ranges. However, the rocks that make up Mt Sonder are up to 900 million years old.

Weathering over the past 300 million years has given the huge granite boulders of the Devils Marbles (above) their familiar rounded shape and left some balancing in precarious-looking heaps. Debris from similar erosion elsewhere in the Red Centre forms the vast parallel ridges of the Simpson Desert (opposite), some of which are 200 km long.

The rocks that form Kings Canyon suffered major fractures during the Alice Springs Orogeny. Layers of Carmichael and Mereenie sandstones, formed 100 million years earlier beneath the vast Amadeus Sea that covered the south-east of the region, were faulted and folded by the upheavals. The upper layer of Mereenie sandstone, stained red by iron oxide, is harder than the Carmichael sandstone beneath it and, as the softer rock has been eroded, the unsupported harder rock has broken away cleanly along vertical fault lines and fallen into the canyon below.

After the Red Centre stabilised about 310 million years ago, there followed a major period of erosion and climate change that resulted in the weathering of the landforms into their present shapes. The granite blocks of the Devils Marbles and the sandstone features of Chambers Pillar, Rainbow Valley and the many mesas south-east of Alice Springs were also eroded during this time and gradually stained red by iron oxide deposits that dissolved in water and leached into the rock. The debris from much of this erosion formed the vast sand dunes and sand plains that dominate the southern part of the region today.

While all this geological activity was occurring, Australia was still part of the super-continent Gondwana and was drifting from Antarctic latitudes to the Equator and south again. About 65 million years ago, when Australia was beginning to separate from Antarctica, Alice Springs was on latitude 45°, further south than Tasmania is today. There were large patches of rainforest, remnants of which can still be seen in Palm Valley and Kings Canyon.

Stately red cabbage palms (above) in Palm Valley are remnants of a rainforest that flourished 65 million years ago. Nearby Kalarranga Lookout (above left) was sculpted by the meanderings of the ancient Finke River. Throughout the Red Centre, right to its southern margins in the Musgrave Ranges (opposite), rivers have carved deep gashes through seemingly impregnable ranges.

Aboriginal History
A 30,000-year heritage

Aborigines have lived in the Red Centre for at least 30,000 years, relying on its landscape for food, water and shelter. Among the many protein-rich creatures they hunted and cooked on hot coals were goannas, which were particularly valuable during periods of drought when kangaroos, wallabies and emus were hard to find. The desert Aborigines also regarded some 140 plant species as food and while the men hunted game, the women gathered seasonal bush fruits and vegetables that were

A vast natural rock shelter in the heart of dune country some 300 km west of Alice Springs is providing a unique window on 30,000 years of Aboriginal occupation.

Until earlier this century, up to 1500 generations of western desert Aborigines used Puritjarra shelter, which nestles in a rocky escarpment in the Cleland Hills overlooking a sandy landscape dotted with spinifex and desert oaks. They decorated it with hand stencils and rock engravings, and the camp-site debris they left behind has helped build up a picture of Aboriginal life in one of the harshest regions of the continent.

Mike Smith, of The Australian National University, has been studying the shelter since 1986. Buried in the sandy floor are stone artefacts, fragments of red ochre, charcoal from fires, animal bones and grindstones – some dating back 30,000 years.

The shelter was used intermittently by family groups from the Kukatja language group (often called a tribe) for so long because it has its own permanent, spring-fed waterhole, shaded by river red gums and bean trees, a little higher up the escarpment.

Like all the Red Centre's Aborigines, the Kukatja moved over large areas within their own country – particularly during the drier times that coincided with the peak of the last ice age, some 18,000 years ago, when they may have had to travel up to 50 km between permanent waterholes or springs.

Their use of the shelter, which is about 40 m wide, 20 m deep and 12 m high, intensified about 7500 years ago, when the climate became warmer and wetter, and became even more frequent 1000 years ago when the population is believed to have increased.

Archaeological evidence indicates that the Kukatja hunted and ate medium-sized marsupials like rat-kangaroos and that about 3000 years ago they began grinding seeds to supplement their diet of game, fruit and vegetables. Their hunting tools also became more sophisticated, with spinifex resin being used to bond stone cutting tools to wooden handles.

The Kukatja were one of more than 20 Aboriginal language groups living a highly structured life in the Red Centre at the time of European settlement. These groups, the best known of which include the Arrernte (often spelt Aranda) from the ranges near Alice Springs, the Warlpiri from the

rich in vitamins. Ironically, the first Europeans in the Centre often suffered from scurvy and other diseases caused by vitamin deficiencies because they ignored – or refused to eat – the abundant bush tucker, even when they were near starvation. Despite the advent of supermarkets and convenience stores, many Aborigines still supplement their diets with bush delicacies like succulent witchetty grubs, which are now finding their way onto many Australian restaurant menus.

MUSEUM OF VICTORIA COUNCIL, COURTESY: ARRERNTE PEOPLE

Ceremonies have always been an important part of Aboriginal life. In 1901 biologist and anthropologist Baldwin Spencer watched this dance (above), which he dubbed the "entrance of the strangers", in Alice Springs during an Arrernte welcoming ceremony. The 30 or so strangers were passing through foreign territory when they encountered a camp. They waited some distance away until two old men approached them and invited them to the camp, then – brandishing spears, woomeras and boomerangs – they danced to meet their hosts. Today, tourists are able to share the spectacle of some ceremonies, like this colourful corroboree (opposite) being performed at Yulara by Stanley Douglas.

north-west and the Pitjantjatjara from the south-west, each had their own recognised stretch of country and were divided into clans consisting of a number of family groups.

The men hunted large game like kangaroos, wallabies and emus with spears and boomerangs, while the women dug for tubers and delicacies such as witchetty grubs and honey ants with digging sticks, and collected fruit and seeds in the wooden containers they always carried. Both men and women hunted lizards and small marsupials. The children usually accompanied the women, but boys spent more time with the men as they neared puberty.

Game was lightly cooked in hot coals and eaten while still blood red. The hunters, elders, initiates, pregnant women and children were each allocated certain parts of the animal and the remainder was shared among the other members of the family group or groups. Seeds were ground and mixed with water to make a paste that was cooked in coals.

It was a varied diet and a healthy one. The game was lower in fat and cholesterol than the domesticated animals eaten by Europeans and the bush fruits and vegetables provided plentiful vitamins.

Although the Aborigines of the Red Centre relied directly on the land for their food, water and shelter, it was only during harsh times that food gathering occupied the greater part of their daily life. Like all of Australia's first inhabitants, they spent much of their time developing and enriching their cultural life. Every aspect of life was governed by rules that were passed on to each generation in stories, songs and art.

Ceremonies were an important part of life, marking events like changing seasons, initiation and childbirth, and they often brought together groups of people from hundreds of kilometres apart for several days of dancing, singing and rituals. At these gatherings, gifts of ochre, artefacts, food cakes and pituri (a leaf chewed as a mild narcotic) would be exchanged.

Early European contact

The estimated 10,000 Aborigines living in the Red Centre in 1860, when John McDouall Stuart made the first of his three attempts to cross the continent, are thought to have been recovering from the devastating effects of a smallpox epidemic. It may have claimed up to 5000 lives – but its impact was minimal compared with the upheavals Aborigines were to face over the next 130 years.

Not surprisingly, the desert Aborigines were alarmed and frightened by the sight of fully clothed white men leading heavily laden horses through their country, and Stuart's diary captures this. Writing of an encounter with "a native and his lubra" at a waterhole on the Hugh River in 1861, he noted: "They appeared to be fishing, and did not see us until I called to them. The female was the first who left the water; she ran to the bank, took up her child, and made for a tree, up which she climbed, pushing her young one up before her."

An entry the previous year commented on the Aborigines' natural, and understand-

able, curiosity. "We were visited by two natives, who presented us with four opossums and a number of small birds and parrots," Stuart wrote. "They were much frightened at first, but after a short time became very bold, and, coming to our camp, wanted to steal everything they could lay their fingers on … They wished to pry into everything, until I lost all patience and ordered them off."

By the time Ernest Giles explored the Red Centre in 1872, Europeans were less of a novelty because Overland Telegraph Line construction teams had been through the area during the previous two years. Giles made an interesting observation after an encounter with a large group of Aborigines at Kings Creek. "I cracked my whip at them – they all made a sudden pause, with a sudden shout, still they did not seem at all inclined to depart. I then unstrapped my gun from my saddle; holding it up I warned them away, and to my great astonishment away they all went," he wrote. "I came to the conclusion, that they had heard from some other half-civilised natives, who had visited the telegraph line, that unless a white man had a gun, he was no more dangerous than an emu or a kangaroo."

The arrival of the first Europeans in the Red Centre after 1860 had a marked impact on Aboriginal life, yet many clung to their traditional way of life until the 1930s. Baldwin Spencer photographed this family group (below) outside their brushwood shelter in Alice Springs in 1896. The man's elaborate body scarring would have been inflicted during initiation, the ceremony that marks the transition from childhood to adulthood. His nose was probably pierced then too, and at least one of his front teeth knocked out. Other Aborigines, like these women (opposite) photographed in 1895, were quicker to adopt aspects of European life, including clothing.

MUSEUM OF VICTORIA COUNCIL, COURTESY: ARRERNTE PEOPLE

The earliest recorded serious conflict between Aborigines and Europeans in the Red Centre occurred in 1874 at Barrow Creek Telegraph Station, when Aborigines fatally speared officer-in-charge James Stapleton and linesman John Franks, and wounded Stapleton's assistant, Ernest Flint. Scores of Aborigines are believed to have been shot in reprisal.

Clashes and conflict

The uneasy relationship between Aborigines and Europeans deteriorated rapidly after the first Europeans settled in the Red Centre in 1872. Two years later, a bloody confrontation at Barrow Creek Telegraph Station, 242 km north of Alice Springs – when scores of Aborigines are believed to have been shot in reprisal for killing two staff members and wounding a third – signalled the start of more than 50 years of bitter conflict.

Not all the European settlers were hostile. In 1877 German missionaries Hermann Kempe and Wilhelm Schwarz arrived in Hermannsburg, 132 km south-west of Alice Springs, having spent nearly 20 months trekking from near Adelaide. The aim of their epic journey was to "civilise" the Aborigines through the gospel and prepare them for assimilation into white society.

The missionaries recorded: "With all the shooting that is taking place, it is hard to conceive that the native people have any kind of future." Against all the odds, the Lutheran mission they established helped to ensure that the Aborigines did indeed have a future – largely because it provided a safe haven in the 1880s and 1890s, when the indiscriminate shooting of Aborigines was widespread.

Elsewhere, though, pastoralists were claiming the best land and tensions were increasing. Not only did the pastoralists dislocate Aborigines from traditional food-gathering areas and sacred sites, but their cattle trampled food plants and fouled water-holes. By the late 1800s, lonely police outposts like those at Boggy Hole and Illamurta Springs (see Chapter 10) were established to try to keep the peace.

As recently as 1928, a tragic confrontation occurred at Coniston station, 232 km north-west of Alice Springs, when dingo trapper Frederick Brooks was fatally speared for allegedly interfering with an Aboriginal woman. Official figures show that 31 Aborigines were killed in reprisal, but the death toll probably exceeded 100.

Despite the tensions, many Aborigines worked on cattle stations as stockmen and

Established in 1877 and now a popular stopping-off point for tourists, the Lutheran mission (above) at Hermannsburg provided a sanctuary for Aborigines in the late 1800s when indiscriminate shooting was widespread. The mission, which eventually had a thriving cattle station, forge, camel-train terminal and tannery to provide employment for some of its 500 residents, was handed back to Aborigines in 1982.

Fire was important to the Aborigines for cooking food, for warmth and for managing the land. They lit small fires to create a patchwork of new growth, on which the animals they hunted could feed, and established growth, in which animals could shelter. Firelighting (above left) was a male skill achieved by rubbing a hard timber, like a mulga woomera, over a groove in softer wood. Friction caused the fine sawdust to smoulder, usually in less than a minute, and this would be blown gently to ignite some tinder.

domestic servants. Pastoralists paid them less than Europeans, if at all, and although they were supposed to provide food, clothing and shelter for their families – with government money – some spent less than they received.

The treatment of Aborigines as second-class citizens permeated all aspects of their lives. No longer able to live off the land, most had to rely on government rations and from the 1930s many were forced to live in supervised camps, missions and government reserves. All were denied basic rights including citizenship, welfare, education and job opportunities, and they were not even allowed into the growing township of Alice Springs unsupervised.

New beginnings

World War II signalled the start of improvements in relations between Aborigines and Europeans that have continued to gain momentum. Some 150 Aborigines played a key role as labourers for the Army in Alice Springs, and later the government developed a network of settlements – with schooling and medical services – that became the basis of the assimilation policy, adopted in 1951, whereby Aborigines were expected to live a European lifestyle.

Many evolved from remote ration posts like Iwupataka, Papunya and Areyonga, to the west of Alice Springs, and although the government's intentions were good, people of different groups were often thrown together in large numbers. This increased tensions and forced them to rely increasingly on European foods because the land couldn't support big populations. Many turned to alcohol to escape their misery.

There were other changes too. Improved road transport put an end to droving in the early 1960s and the number of Aborigines employed on cattle stations fell from 427 in 1958 to 221 in 1965. Numbers fell further in the 1960s when Aborigines won the right to equal wages.

In 1974, two organisations that were to have a major impact on Aboriginal life were formed. One was the Central Land Council (CLC), set up to investigate land rights; the other was Tangentyere Council, which pressed for housing leases for Aborigines wanting to live in Alice Springs. Within 10 years it had established 15 town "camps", usually of brick-veneer houses, accommodating some 1500 Aborigines.

The CLC became a powerful champion of Aborigines after the 1976 Aboriginal Land Rights Act. Its 83 members represent traditional landowners and assist with land claims, sacred sites protection, development, land management and permits.

Aboriginal resurgence

The resilience and resourcefulness of the Red Centre's Aborigines have helped them to weather the turbulent 130 years since European settlement, and their culture and traditions are now enjoying something of a resurgence – thanks largely to tourism.

Kit Whillock, trainer/coordinator of Kurkara Tours, based at Kings Canyon, explained: "The elders see tourism as their children's future." Tourism not only provides

The distinctive dot paintings of the western desert Aborigines like Pinta Pinta (above) and Thomas Rice and Jeannie Nungarrayi (opposite) depict Dreaming stories. The dots are a modern – and more lasting – representation of traditional sand "paintings" in which seeds, flower parts and feathers were used to decorate ceremonial grounds for special occasions.

An Aboriginal artist painstakingly applies acrylic paint dots to his canvas (above). Works like this one (right), displayed by Mick Namarari, 1994 winner of the prestigious $50,000 Red Ochre award for prominence in the Aboriginal art world, are in great demand – and command high prices – in Australia and overseas. Mick is one of more than 250 artists and craftspeople who sell their work through the Papunya Tula cooperative in Alice Springs. Founded in 1971 in Papunya, 240 km north-west of Alice Springs, the company now extends its operations into Western Australia because many of its artists have moved back to their traditional lands there.

Nellie Patterson (left) uses a hot wire to decorate a carved snake at the Maruku Arts and Crafts Centre at Uluru. Crafts, mostly representing desert wildlife and Aboriginal tools and weapons, are collected from 15 remote communities by truck and sorted and oiled at Maruku's warehouse before being distributed to outlets throughout Australia. Income from sales is ploughed back into buying more crafts and providing craftspeople with tools and materials. FLYING ANT DREAMING (below) is one of the many paintings by well-known Yuendemu artist Clarise Nampijinpa Poulson, who sells her work through Papunya Tula.

valuable income but it also encourages Aborigines, who make up almost a third of the Red Centre's population of 36,000, to keep their culture alive.

As well as working directly in tourism, hundreds of Aborigines make a living by selling their art and craft work. Most artists use the western desert dot style to depict their traditional ceremonies and stories in acrylic on canvas and other surfaces. Two large cooperative ventures, Maruku at Uluru (see Chapter 10) and Papunya Tula, based in Alice Springs, each sell the work of up to 800 talented wood carvers and artists, and there are many smaller outlets throughout the region.

Another factor that's helping to ensure Aborigines retain their traditions is the growth of the outstation movement. The first outstations – where Aborigines follow a simpler, more traditional lifestyle than in Alice Springs and the communities scattered throughout the region – were established in the late 1940s, but the movement has flourished in the past 20 years.

Glen Auricht, manager of the Tjuwanpa Resource Centre at Hermannsburg, helps keep 33 outstations, housing 600 people in a 6000 sq. km area, ticking over. "We work 12 hours a day, six or seven days a week," said Glen, who heads 20 staff and Aboriginal workers who do all the accounting and administration for the outstations, as well as providing construction and garage workshops, and other practical resources. Glen and the staff are directly responsible to their Aboriginal management council of traditional owners.

Tjuwanpa Resource Centre at Hermannsburg helps Aborigines to follow a more traditional, less urbanised, lifestyle on 33 outstations by providing administrative support and workshop and other facilities. In the bustling workshop, Mark Inkamala (right) welds a feral cat trap and Frank Fejo (top) repairs a chainsaw. The cat traps are one of many projects through which Aborigines earn their welfare payments under the Community Development Employment Program. It's a scheme the Aborigines themselves voted to adopt and Glen Auricht, Tjuwanpa's manager, believes it works well. Mark Fly (above) helps to administer the scheme from the chaotic office he shares with Glen, while Conrad Ratara catches up on paperwork.

Aborigines wanting to establish an outstation on their traditional land first have to prove their commitment by camping there for at least two years. Glen says it's a tough life. "We encourage people to work hard to be independent," he said. "They do a lot of hunting for traditional food and it is to their credit that they live so well with the land."

Like all the Red Centre's Aborigines, those on outstations earn their unemployment benefit by working at least 20 hours a week – on community projects that range from building and roadwork to growing food, managing feral animals and childcare – under the Community Development Employment Program.

But perhaps the biggest change to affect Aborigines is that they are now major landowners. The handing back of Uluru and Kata Tjuta to their traditional Pitjantjatjara owners in 1985 symbolised the growing recognition of the rights of the first Australians, and Aborigines now own about half of the land in the Red Centre, including 13 pastoral leases.

Aborigines have regained about half the land in the Red Centre and entry to much of it is by permit only (above). But as some barriers go up, others come down and an increasing number of communities and outstations are welcoming tourists, and inviting guests to share their 30,000-year-old culture. At Cave Hill outstation, 100 km south of Yulara, resident Minnie (left), who like many Aborigines doesn't have a surname, displays protein-rich witchetty grubs dug from the roots of the witchetty bush. Yulara-based Desert Tracks takes small tour groups to Cave Hill and nearby Angatja for up to a week. Visitors live simply with their hosts – joining them on bush tucker expeditions, listening to oral history and stories, and visiting sacred sites with them.

The pioneers of the Red Centre

Rising more than 50 m from the surrounding plain, imposing Chambers Pillar was an important navigational beacon for the early European explorers and pioneers. It was named in 1860 by John McDouall Stuart, the first European to venture into the Red Centre, after one of his sponsors. Stuart encountered the distinctive sandstone column during the first of his three attempts to cross

On 22 October 1872, a brief message tapped out in morse code in London crossed the world to Adelaide and heralded the dawn of a new era for the growing colony of Australia.

The message, the first telegraphic communication between England and Australia, was to South Australian Governor Sir James Fergusson and read: "As chairman of the British-Australian Telegraph Company, I congratulate you and the colonies of Australia on the completion of a great work. The spirit and determination of South Australia have nobly combined with the enterprise of the Mother Country in triumphing over all our difficulties."

It was from Lord Monck, the chairman of the company that effectively ended Australia's isolation from the rest of the world. His company's ambitious scheme to link a submarine cable from Java – already connected to Europe – with an overland telegraph line that stretched 2900 km across Australia meant news could now cross the world in a few hours instead of taking at least three months by sea. And it was all made possible by the determination of one man – John McDouall Stuart, our greatest inland explorer.

His resolve to become the first European to

cross the continent from south to north paved the way for the telegraph line that ran from Port Augusta in the south to Port Darwin (now Darwin) in the north. It also made him the first European to venture into the Red Centre – in 1860, an incredible 72 years after colonisation.

Curiosity about what lay in the centre of the continent had been intense for many years. Charles Sturt, often regarded as the father of Australian exploration, believed it contained a vast inland sea and tried in vain to prove this on an 1844–46 expedition. Among his party was Stuart, a capable Scot who, on 2 March 1860, embarked on the first of three attempts to cross Australia from south to north.

Stuart's expedition was a modest one. Accompanied by William Kekwick and Benjamin Head, he left Stuart Creek in South Australia with 13 horses and provisions for six months. He headed north-west to the Finke River, whose waterholes were unusually plentiful after good rain, and Chambers Pillar, which he named after his patrons, William Finke and James Chambers.

On 12 April, he reached the MacDonnell Ranges and noted: "The country in the ranges is as fine a pastoral hill-country as a man would

Australia from south to north to find a suitable route for an Overland Telegraph Line to link the growing colony with "Mother England". Not surprisingly, it became an important landmark for other explorers, including Ernest Giles, and for the early settlers. Many etched their names into the soft rock near the base of the pillar, which has survived because its hard capping has resisted erosion.

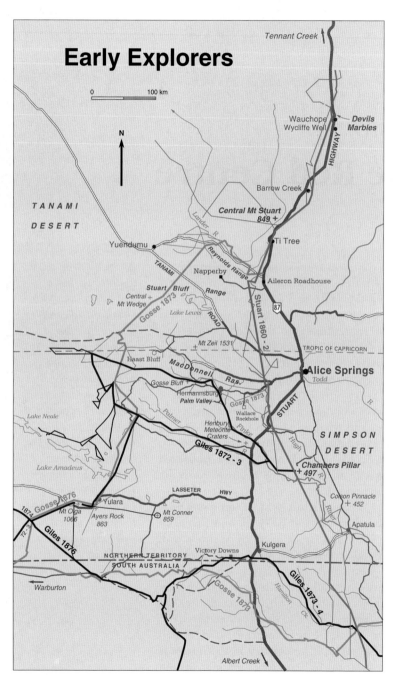

Early Explorers

wish to possess; grass to the top of the hills, and abundance of water through the whole of the ranges."

He continued north to what is now Central Mount Stuart, which he calculated to be 4 km south-south-west of the geographical centre of Australia, then north-west into the Tanami Desert, where water shortages and health problems forced him to retreat to near Central Mount Stuart. On 16 May he wrote: "I have been suffering dreadfully during the past three weeks from pains in the muscles, caused by the scurvy, but the last two nights have been most excruciating. Violent pains darted at intervals through my whole body. My powers of endurance were so severely tested, that, last night, I almost wished that death would come and relieve me from my fearful torture." His hands were a mass of sores and his mouth and gums were so painful that he even had difficulty swallowing the flour and boiled water that had become his staple diet.

When he felt a little better, the trio continued to Tennant Creek but three weeks later, on 27 June, they were turned back 70 km further north after being challenged by hostile Aborigines at Attack Creek – just 600 km from their goal.

Stuart arrived back in Adelaide on 7 October and almost immediately the South Australian Parliament voted a sum of £2500 for a larger and better-equipped expedition. Explorers Robert O'Hara Burke and William Wills had left Melbourne on 20 August to cross the continent, and the South Australians didn't want the Victorians to pip them at the post.

On 1 January 1861 Stuart set off with his second-in-command William Kekwick, 10 other men and 49 horses. They reached Newcastle Waters, but a shortage of supplies forced them to retreat on 11 July. Stuart wrote: "We are all nearly naked, the scrub has been so severe on our clothes … The men are also failing and showing the effects of short rations."

Stuart reached Adelaide on 23 September in poor health, but agreed to undertake another expedition almost immediately and on 8 January 1862 he set off with 11 men and 71 horses. By this time it was known that Burke and Wills had perished on the way back from their successful crossing of the continent by a more easterly route.

A jubilant Stuart reached his goal on 24 July, but his health deteriorated and by the time he approached the Red Centre on the return journey, he was very sick. On 18 October he wrote: "I am now so helpless that I have to be lifted into the saddle." Ten days later he noted: "I am now reduced to a perfect skeleton, a mere shadow … I can chew nothing, and all that I have been living on is a little beef tea, and a little boiled flour."

His colleagues built a stretcher for him and he lay on that for much of the journey, finally arriving back in Adelaide in December. Despite the hardships he had encountered, he was enthusiastic about the telegraph line. "There would be a few difficulties in the way, but none which could not be overcome and made to repay the cost of such an undertaking," he wrote.

He returned to England and died in London in 1866, well before work started on

Sections of the usually dry Finke River (above), named by Stuart in 1860 after his second sponsor, were a "highway" for the early explorers. "The creek is very large, with the finest gum-trees we have yet seen, all sizes and heights," Stuart noted in his diary. "I have not passed through such splendid country since I have been in the colony." He also noted the distinctive mesas that dot the landscape to the south-east of Alice Springs. Less than three weeks after leaving Chambers Pillar, Stuart and his second-in-command William Kekwick climbed Central Mount Stuart (right), which they calculated to be within 4 km of the geographical centre of Australia.

the line. But his careful records of permanent water and suitable timber meant that when the line was built, it mostly followed his route. So too did the railway and, finally, much of the Stuart Highway linking Adelaide, Alice Springs and Darwin.

The Overland Telegraph Line

In June 1870, Charles Todd, South Australia's Superintendent of Telegraphs, was given the awesome responsibility of overseeing the construction of the Overland Telegraph Line. It was Australia's most ambitious engineering feat with two-thirds of the line running through inhospitable country untrod by Europeans until Stuart's expeditions.

Under the terms of the contract with the British-Australian Telegraph Company, some 36,000 poles had to be cut and erected at 80 m intervals across the continent, the single iron wire had to be installed and repeater stations had to be built every 250 km to boost the telegraphic signal before it faded – and all in just 18 months.

Todd divided the route into three sections and judged the most difficult to be the central one from Oodnadatta to Tennant Creek. He engaged private contractors to build the northern and southern ends, but decided to use government staff for the middle.

Survey parties under the leadership of John Ross went ahead of the construction parties to determine the best path for the central section. They largely followed Stuart's route, but found his passage through the MacDonnell Ranges too rugged and selected an alternative route. Later, 5 km north of the ranges, a suitable site for a repeater station was found next to a permanent waterhole. The river in which it sat was named the Todd, after Charles Todd, and the waterhole was named Alice Spring for his wife.

Despite the hostile terrain, the searing summer heat and the erratic water supplies, the central section was finished on schedule in December 1871 and the following month the first telegraphic message was transmitted to Adelaide. Delays to the northern section and problems with the submarine cable that brought the line ashore at Darwin meant the first overseas communication was not made until the following October, but Charles Todd and his men were rightly feted as heroes. Within a year of the line opening, 4000 telegrams had been transmitted, Australia was no longer isolated from the rest of the world and Europeans were settling in the Red Centre.

Putting the Centre on the map

Stuart may have paved the way for the Overland Telegraph Line, but it was Ernest Giles, the most eloquent of the great explorers, who put many of the Red Centre's attractions – including Finke Gorge National Park, the Olgas, now known by their Aboriginal name, Kata Tjuta, and Kings Canyon – on the map and described them so lyrically in his diaries.

In August 1872 he led an expedition to look for pastoral country to the west of the telegraph line and to try to find a route to the west coast. Accompanied by Samuel Carmichael, Alexander Robinson, 15 horses and a dog, he left Chambers Pillar and

PIP AND DICK SMITH

MORTLOCK LIBRARY, STATE LIBRARY OF SA

*The parallel ridges of the MacDonnell Ranges (above) proved a formidable barrier
to the early explorers. Overland Telegraph Line construction teams found Stuart's route
through the ranges too rugged and settled on one further east, near where Alice Springs
is today. The original timber poles were soon damaged by termites and were replaced
with iron poles (opposite) and a second wire was added in 1898 to cope with the
increased traffic. Soon after the line was completed in 1872, Ernest Giles (above right)
set off on the first of three expeditions to look for pastoral country in the Red Centre
and find a route to the Western Australian coast. Later he was granted a lease on
2525 sq. km of pastoral land, including the whole of what is now Watarrka National Park.*

Lake Amadeus and a network of smaller salt lakes (below) robbed Giles of the honour of naming Ayers Rock. He saw the rock and near-by Mt Olga in 1872, but the salt lakes were a barrier to reaching them until the following year, by which time William Gosse had climbed and named the rock now known by its Aboriginal name, Uluru. Giles was still mapping the Centre when Edward Bagot took up the pastoral lease on Undoolya station (opposite above), on the outskirts of Alice Springs, which was settled in 1873 by James Churchill-Smith. Present owners Gail and Jim Hayes and their son Andrew (opposite below), who are descendants of early pastoralist William Hayes, often thumb through Churchill-Smith's diaries in the kitchen of their 1873-built homestead.

followed the Finke River to the Glen of Palms and the MacDonnell Ranges, where he headed west, then south.

There he encountered the salty expanse of Lake Amadeus. "We walked a long distance on its surface, and to our weight it seemed firm enough, but the instant we tried our horses they almost disappeared out of sight," he noted. It proved an impenetrable barrier to reaching Mount Olga, which he believed was the gateway to better country, so he retreated north to review his plans. There he named Kings Canyon and the George Gill Range, observing: "The country round its foot is the best, which I have seen in the region; and no doubt before many years pass it will be part or all of a stocked run."

It was now November and the intense heat and shortage of food were taking their toll. Giles retreated to reorganise his expedition and set off again the following August with William Tietkens, Alfred Gibson, James Andrews and 24 horses.

An expedition led by William Gosse and equipped with camels had set off for the west in the meantime. Giles, who followed a more southerly route on his second expedition,

believed the other party was travelling further north, but on 14 September, at Mount Olga, he was shocked to see horse, camel and dray tracks. He realised they belonged to Gosse's party and wrote: "Had the earth yawned at my feet, for ever separating me from this mountain, or had another of similar appearance risen suddenly before my eyes, I could not have been more astonished … It was almost a death-blow to my expedition." Gosse named and climbed Ayers Rock – now known by its original name, Uluru – which had been sighted the previous year by Giles, but failed to reach the west coast.

Giles continued and penetrated some 350 km into Western Australia when disaster struck. Gibson was lost in the desert that now bears his name and Giles and the others retreated, devastated by the tragedy. The following year, equipped with camels, Giles finally reached his goal.

Pastoral and urban expansion

While Giles was busy mapping the Red Centre, pastoralists were beginning to take out leases. Undoolya, on the eastern outskirts of Alice Springs, was among the first stations to be established, in 1873, by James Churchill-Smith, with 1500 cattle. Present owners Jim and Gail Hayes still have copies of his diaries in their 1873-built homestead, which they claim is the oldest inhabited building in the Northern Territory.

The Hayes are the region's oldest-established pastoral family. Jim's great-grandfather William settled in the area in 1882, shortly after completing a contract to cart in iron poles for the reconstruction of the Overland Telegraph Line. Aborigines had removed some of the original timber poles to use the porcelain insulators near the top to make cutting tools; others had been damaged by termites.

William settled Deep Well and Maryvale stations, to the south of Alice, and in 1907 he acquired Undoolya, which later passed to his son and grandson, both called Ted, and finally to Jim.

By the late 1800s, most of the land that was even remotely suitable for pastoral use had been snapped up. The pioneer pastoralists found life tough. As well as having to cope with extreme isolation, their land was far from ideal for grazing and water was in short supply. After overlanding cattle – usually from the Murray area of South Australia – to stock their stations, they had a virtually non-existent local market for their meat and faced prohibitive costs in getting cattle to the southern markets.

Hot on the heels of the pastoralists came a new wave of settlers. The chance discovery of gold at Arltunga (see Chapter 9) in 1887 lured hundreds of eager prospectors to the east MacDonnell Ranges and the following year the township of Stuart, as Alice Springs was known until 1933, was surveyed to service the growing population.

Growth of the new town, 3 km south of the telegraph station, was slow, despite the flurry of activity in the region. Goods had to be brought by camel train and although a railway between Adelaide and Oodnadatta opened in 1891, travellers faced a circuitous two-week, 600 km buggy ride from the railhead before they reached Stuart. In 1903,

The little town of Stuart, as Alice Springs was called until 1933, received the biggest boost of its short life in 1929 with the coming of the Ghan (right). The train, named after the Afghan camel trains it replaced, effectively ended the isolation of the Red Centre by providing a link with the more-populated south. The importance of that link became evident little more than 10 years later, when Alice Springs became the major military supply base for the war effort further north. The Darwin Overland Maintenance Force established a large camp (below) at the foot of Anzac Hill in 1940, and the first of almost 200,000 troops and 500,000 tonnes of supplies began arriving by railway. Many of the town's facilities were expanded to cope with the influx of people, who travelled north in massive convoys along the hastily improved Stuart Highway. When hostilities ceased, many of the troops returned as tourists for a more leisurely look at the Red Centre.

THE GHAN PRESERVATION SOCIETY

FRANK TRANTER COLLECTION, NORTHERN TERRITORY LIBRARY

when the population of Arltunga and nearby Winnecke peaked at about 400, Stuart had just 35 non-Aboriginal residents and this barely doubled over the next 25 years.

It was the extension of the railway from Oodnadatta to Stuart in 1929 that gave the Red Centre its biggest boost. The fortnightly service, christened the Ghan because it replaced the Afghan camel trains, revolutionised communications – despite sections of its primitive narrow-gauge track needing frequent repairs when they were washed away by flash floods.

Just 10 years after its completion, the railway that had been built primarily to advance the population and development of the region took on a major new role, when Stuart – now Alice Springs – was catapulted to strategic importance during World War II.

World War II

The 900 people living in Alice Springs at the start of World War II in September 1939 could never have imagined that, over the next five years, the Ghan would bring almost 200,000 troops and 500,000 tonnes of supplies to the town. They were destined for the war effort in Darwin, 1300 km to the north.

Suddenly the little town that existed mainly to service the outlying pastoral properties experienced a population explosion as it became the headquarters of the Darwin Overland Maintenance Force. Military tent cities were established and the population swelled further when the town temporarily replaced Darwin as the administrative centre for the Northern Territory.

War correspondent George Johnston noted in a 1941 newspaper article: "The streets of the prettiest town of the inland are thronged each night with men in khaki … At Alice Springs you will see anything from 500–1000 huge Army trucks parked in regular lines waiting for orders to head north. And on the other side of the hill you will see the camps – the neatest, cleanest, and most efficient camps I have seen anywhere in Australia."

The Stuart Highway was sealed north of Alice Springs to make the journey a little less arduous for the hundreds of convoys, power and water supplies were boosted and aerodrome, hospital and recreational facilities were expanded to cater for the influx of people.

By the time the war ended in 1945, Alice Springs had changed forever. And it was heading for the biggest expansion in its short life because the thousands of troops who passed through had glimpsed some of the region's abundant natural attractions. They were impressed – and it wasn't long before some returned as tourists, laying the foundations of a burgeoning $200 million industry.

COURTESY: JIM COTTERILL

The growth of tourism

Jim Cotterill leafed through a pile of fading photographs, pausing occasionally for a closer look at one. "These were the old Army blitzwagons that took the first tourists to Palm Valley," he said, pointing at a vehicle that looked like a cross between a coach and a tank. "They were pretty uncomfortable. You'd sit there rugged up with earmuffs, beanies, scarves and blankets and as the day warmed up you'd throw the clothing off."

Jim first travelled to Palm Valley, 125 km south-west of Alice Springs, as a 9-year-old in 1953, after his father Jack started working for tourism pioneer Len Tuit, who drove tourists there from the late 1940s. The journey, which now takes just over two hours, took a full day.

Soon afterwards, the young schoolboy joined one of the early trips to Ayers Rock. "It took two days to get there and there were dirt roads all the way," Jim said. "In the sandhill country you got bogged more times than you didn't. The passengers would get out and collect spinifex and wood to corduroy the wheel tracks to form a firm base. They'd then get behind the bus and push – it was a definite adventure."

The early tourists to Ayers Rock travelled by bus and camped at the rock, but in 1958 Jack built permanent accommodation and joined forces with Eddie Connellan, founder of the Centre's first airline, to fly tourists in and out.

The accommodation – a corrugated-iron shed – slept 12 and was a far cry from today's luxury air-conditioned resorts. But it was a success and the following year a similar shed, sleeping 16, was built at Palm Valley. Tourist facilities were also established at Glen Helen and Serpentine Chalet in the west MacDonnell Ranges, and at Ross River in the east MacDonnells.

In 1960, Arthur Liddle of Angas Downs station told Jack about a canyon 253 km

JIM COTTERILL

The first organised tours to Ayers Rock were a far cry from those of today. Adventurers on this 1951 trip (top) camped at the base of the rock after a two-day journey from Alice Springs in a no-frills vehicle similar to the Chevrolet Blitzwagon (above), pictured negotiating the rocky terrain at Palm Valley. By the late 1950s, tourist numbers were increasing so rapidly that more permanent accommodation was built at a number of destinations.

south-west of Alice Springs. "He said there were waterholes, sheer cliffs and palm trees," Jim said. "I remember my father saying that if it's half as good as Arthur reckons, it would have to be the best place in central Australia." Jack and Jim visited the canyon – Kings Canyon – and decided to open it up to tourists.

Arthur offered Jack land near Yowa Bore to build accommodation and helped Jack and Jim cut a 103 km track to the canyon. "Most of the work was done in the summer of 1960–61, at the hottest time of year," Jim said. "We chopped the trees down, dug out the roots and cleared an area by hand, then we'd walk back and get the truck and tow a steel drag up and down a few times."

In May 1961 they started the first trips from Alice Springs to Kings Canyon, attracting 250 patrons on three-day tours in the first season. The venture flourished and by the time their base – Wallara Ranch – closed in 1990, following a disagreement over the lease, it was catering for 30,000 people a year.

Meanwhile tourism to Ayers Rock was escalating and in 1974 the Northern Territory Government decided to investigate the development of a resort town on a site 12 km north of the rock. Ten years later Yulara became fully operational and it now caters for more than 300,000 visitors annually.

In little more than 100 years, Europeans have changed the face of the thriving Red Centre forever.

By the late 1960s and early 1970s, the burgeoning tourist facilities near the base of Ayers Rock were becoming so unsightly that the Northern Territory Government was prompted to create a proper tourist infrastructure away from the rock now called Uluru (above). The ultra-modern resort town of Yulara opened in 1984 on a site 12 km north of the rock, and road improvements since then mean visitors can travel between Yulara, Uluru and Kata Tjuta on sealed roads. More recently, road upgrading and a resort development near Kings Canyon (opposite), which wasn't opened up to tourists until 1961, have made Watarrka National Park much more accessible. What hasn't changed, of course, are the natural splendours that the growing number of visitors come to see.

The Bustling Centre

Cattle auctions are a regular feature of the Alice Springs calendar and attract strong interest from buyers all over Australia and, increasingly, from overseas. At an auction run by Elders at the Bohning stockyards, auctioneer and Alice Springs branch manager Arthur Hansberry, centre in brown jacket, appraises stock with potential bidders. Up to 2500 head from

"Good morning, Mr France," 10 voices chanted in unison from all over a 1.3 million sq. km classroom. It was Wednesday at 10.50 a.m. and Year 5 teacher Ted France was behind the microphone in the studio of the School of the Air in Alice Springs, greeting his 10-year-old pupils.

"I've got some news," volunteered an enthusiastic Kylie from Suplejack station, 650 km northwest of Alice Springs. "On Friday night we had 18 mm of rain. We haven't been mustering much since, because it dampened the ground."

It was the sort of chatter that goes on in country classrooms throughout Australia, but these were kids who meet each other only twice a year, and their teacher just three times a year. The rest of the time their only contact with their classmates is through their three half-hour radio lessons each week with Australia's first School of the Air, which was established in 1951.

Ted moved on to maths, keeping the pace lively to maintain interest. He asked each pupil questions that involved squaring and cubing figures – then asked them to explain how they'd arrived at their answers. "It's really a question of how well we know our multiplication tables," he said.

Then there was time to read a pupil's poem, to

thank another for the furry toy she'd made and for a few final messages before he said "cheerio" to his class.

On the way back to his desk in the open-plan office he shares with the school's 10 other teachers, Ted stopped at a noticeboard displaying photos of his pupils and pointed out where they live – on remote cattle stations and in roadhouses, mining and Aboriginal communities – on a map.

He's visited every one of them at their home, always staying overnight, and it was evident that he probably knows them and their families better than many city teachers know the kids they see every day.

In addition to their three lessons, pupils have a library session, when they listen to a story, and a 10-minute personal chat with their teacher each week to discuss their progress. The rest of the time they work their way through correspondence lessons.

Back at his desk, surrounded by the bright paintings and craft work that adorn most classroom walls, Ted explained that the school's 120 pupils – who range from pre-schoolers to Year 7 – come to Alice Springs twice a year for a few days to get to know their classmates and their teachers, and that teachers visit each of their pupils once a year.

stations in the Alice Springs area are sold at each auction and some are exported live to Asia, where they are highly prized. Once the lifeblood of the Red Centre, the cattle industry has been overtaken by tourism in the past 30 years but still contributes about $30 million a year to the local economy and provides employment for an estimated 400 people.

Year 5 pupils Emma Petrie, left, and Holly Delves enjoy the unusual experience of sitting in on a School of the Air lesson with their teacher, Ted France. The girls were visiting Alice Springs from their homes in Kintore, 450 km to the west, to say goodbye to their sisters, who were heading off to Tasmania on a school trip, and paid Ted a surprise visit. The school has welcomed thousands of visitors – including the Prince and Princess of Wales, former British Prime Minister Margaret Thatcher, and a host of stage, screen and sports personalities – and many have autographed the studio wall.

"I taught in the bush and in Alice Springs before joining School of the Air 14 years ago," he said. "I love this job because it gives me the best of both worlds."

Although most lessons are conducted by two-way radio, the school also beams some televised lessons to pupils by satellite and makes use of the phone and fax, relatively recent additions to many remote properties. "In 14 years we've come from the 'horse and buggy' days to the latest in technology," he said.

Population and employment

Advancements in technology and communications have greatly changed life everywhere in the past 50 years – and perhaps nowhere more dramatically than in the Red Centre. New industries, including tourism, gas and oil exploration and defence, have brought an influx of people – and created a booming economy with unemployment at a low 6 per cent, about half the national average.

The population of Alice Springs, home to more than 25,000 of the region's 36,000 inhabitants, has grown fivefold in the past 30 years and it's become a thriving town with services and facilities to rival those of cities twice its size. Up to half of its residents, including many of the 4000 Commonwealth, Territory and local government staff and most of the 800 defence staff, are on three- to five-year postings, so there's a constant turnover in population and a youthful vibrancy to the town.

(Note: When government departments define the Red Centre, they refer to a 540,000 sq. km area covering the southern half of the Territory from Tennant Creek and stretching to the WA, SA and Queensland borders. Figures quoted in this chapter are from government sources so do not correspond exactly with the smaller 300,000 sq. km area covered by this book.)

Primary industry

Until 30 years ago, the pastoral industry was the economic backbone of the Red Centre. Today, although it uses nearly half the land, it has been overtaken by tourism. But it's still a significant industry, with cattle alone earning an estimated $30 million a year. The 86 cattle stations in the region support about 300,000 head – mostly Herefords and shorthorns – but these figures fluctuate depending on rainfall and beef prices.

Pastoral employment has dropped significantly in the past 30 years to about 400. Jim Hayes, the fourth generation of the Hayes family to own the lease on Undoolya, a 1490 sq. km property with 5000 poll Herefords on the eastern outskirts of Alice Springs, once employed eight to 10 people. Now just he, his wife Gail and their three sons work there. "With motorbikes for mustering and cattle trapped in yards, we've become so efficient that we can do 12 months' work in about three months," he said. "The other nine months become a problem because whatever you do costs money."

Some owners have diversified into tourism and other activities to provide more work. Ian and Lyn Conway, who own 1100 sq. km Kings Creek station, 230 km south-west of

Boot-scooting country music enthusiasts Brett Heaslip and Tanya and Troy Dann (above) can sing about long days in the saddle with conviction because they all work on cattle stations and have tasted the dust and the flies. Their eight-piece group, the Outback Legends, has entertained audiences as far afield as the Tamworth festival in NSW and so impressed singer Lee Kernaghan that he invited them to take part in his film, BOYS FROM THE BUSH.

Alice Springs, have 800 breeding cattle but regard tourism and camel catching as their major money earners.

Their attractive camping ground caters for up to 400 people a night and many more call in for petrol or supplies on their way to nearby Watarrka National Park (see Chapter 10). But it's camels – imported from 1840 to open up outback Australia and left to run wild when motorised transport and railways made them redundant in the 1920s – that are the major growth industry.

Over the past 25 years, Ian has caught thousands of the 500 kg beasts for camel farms and for live export to the USA and the Middle East. More recently, many have been destined for the table.

"Camels get a bad press but nobody sees the good side of them," Ian said. "They're a valuable asset producing $2.6 million towards the Territory budget. By the turn of the century it will be a $10 million industry."

Despite its arid climate, the region also supports a small but significant horticultural industry. Table grapes, mangoes, asparagus, figs and citrus are grown near Ti Tree, 185 km north of Alice Springs, along with kangaroo paw and other native flowers that are exported to Japan, netting more than $100,000. The crops are irrigated with water from a 5000 sq. km subartesian basin (see Chapter 12).

Tourism

The most visible industry in the Red Centre is tourism, which brings more than $200 million a year to the local economy and provides employment – either directly or indirectly – for about 8000 people. In 1993–94, 494,000 people visited the region, a 5.3 per cent increase over the previous year, and spent $228 million.

With their school friends often hundreds of kilometres away, outback kids like Sam and Georgina Chisholm (above) become adept at making their own entertainment. When 18 months of drought at Napperby station, 175 km north-west of Alice Springs, came to an end, the youngsters celebrated by heading for the nearest puddle. In drier conditions at Amburla Brahman Stud (left), 100 km north-west of Alice Springs, Troy Dann sends the dust flying as he takes off on his motorbike to muster cattle for live shipment to Asia. Troy's music and film production activities and his management of the stud, which runs 2000 Brahmans, earned him the Northern Territory's Young Achiever award in 1994. Although he uses a motorbike whenever possible because it saves time, he also musters on horseback (opposite).

Production operator Gary McFadzean checks a gas reinjection well at the Mereenie oil and gas field (above). Surplus gas is put back into the ground, which helps to stimulate the flow of oil at the field, one of two to the south-west of Alice Springs that produces gas to generate most of the Territory's electricity. Later, Gary pauses beside one of the process flares (above right) that are part of the field's safety system.

Northern Territory Tourist Commission figures show that 98,000 visitors were from the Territory, 189,000 from interstate and 203,000 from overseas. The average length of stay was 4.6 days and daily spending by each visitor averaged $100.

Tourism is set to continue to expand and a number of major attractions are currently being developed. These include the Pioneer Women's Hall of Fame in Alice Springs, aimed at highlighting the role women have played in both urban and rural Australia, and the $24 million Desert Wildlife Park and Botanic Gardens being developed on the outskirts of Alice Springs by the Conservation Commission of the Northern Territory and due to open in 1997. The park will provide visitors with a unique opportunity to discover the different plant and animal habitats of the region and to see the often elusive creatures that live in them.

Energy production

Switch on a light or an airconditioning unit in any major population centre in the Territory – and many smaller ones – and it will be powered by electricity generated by gas discovered in the Amadeus Basin, south-west of Alice Springs, in the mid-1960s.

Natural gas reserves from 2500 m underground at the Palm Valley gas field at Hermannsburg, 132 km south-west of Alice Springs, and Mereenie, 126 km further west, generate electricity for much of the Territory. Some gas is pumped to Alice Springs, but most travels along a 1500 km pipeline to Darwin. About 80 per cent comes from the six wells in production at the Hermannsburg facility, which is staffed by six people, and the rest is from Mereenie, which has a staff of 26. Some 400,000 litres of oil are also extracted every day from the 30 wells in production at Mereenie and are pumped to Alice Springs then transported to Adelaide for refining.

Defence

A cluster of mushroom-like domes at Pine Gap, 18 km south-west of Alice Springs, is the most visible sign of a major US-Australian intelligence centre that became operational in 1970 and brought nearly 400 American families to the Red Centre.

The Joint Defence Facility Pine Gap is a civilian establishment employing up to 800 staff, more than half of whom are Australians, and is a key part of the local economy. It's estimated that in 1993–94, expenditure on salaries was $29 million – much of which was spent locally – and that $18 million went on Australian-sourced supplies and $2.5 million on utilities including housing rentals. Among the sports the Americans are credited with bringing to the town are gridiron and baseball.

Another defence facility, the Jindalee Over-the-Horizon Radar Network, 27 km north-west of Alice Springs, employs about 55 people who monitor air and sea approaches to northern Australia.

The Royal Flying Doctor Service

In the operations room of the Royal Flying Doctor Service (RFDS) at Alice Springs, Steve Byrnes surveyed the white board showing aircraft movements. One plane was on its way to pick up a mother and baby, classified as Code 1 or serious, in Yulara, 335 km to the south-west, another was flying the 205 km to Mt Ebenezer, to a stretcher case classified as Code 2 or less serious, and the third was undergoing routine maintenance.

"If we get another call now, the doctor who takes it will have to decide on priorities," he said. "If it's a Code 1, we could redirect the plane on its way to Mt Ebenezer."

Steve, manager of Northern Territory operations for the RFDS, which was founded in 1928 to provide "a mantle of safety" for people living beyond the reach of conventional medical services, was explaining the role of the Alice Springs base.

"In the past year we flew 446,092 nautical miles (826,162 km) in 2666 hours, transported 2111 patients and took 1952 staff to hold clinics," he said. "We cover an area of about 1 million sq. km and serve 16,000 people outside Alice Springs."

Although the service is best known for its role in evacuating seriously ill and emergency patients – both residents and tourists – to hospital, it also provides medical care and advice through radio and telephone consultations, and operates regular medical and dental clinics in isolated communities.

While Steve outlined the work of the RFDS, radio operator Barry Williams was dealing with a constant stream of radiophone calls, including one from a road-grading crew near Borroloola, 894 km to the north, who were ordering their weekly supplies. He encourages 4WD adventurers equipped with high-frequency radios to ring in daily. "Sometimes the police will contact us if they're worried about someone," he said. "If we've heard from them we can let the police know they're OK."

Barry, who coordinates aircraft movements, explained that until the recent expansion of the outback telephone network, the service's high-frequency radios were the only link many remote properties had with the outside world.

Pine Gap (top), a high-tech US-Australian intelligence facility, is a major employer with about 800 staff, more than half of them Australian. The sensitive nature of its work means visitors are not admitted, but tourists can watch staff like Diana Armstrong (above) plotting aircraft movements at the Royal Flying Doctor Service base in Alice Springs. The base provides essential medical services for 16,000 people in remote communities.

The Living Centre

The Red Centre is home to hundreds of creatures, but most visitors will see only a fraction of them. Many are nocturnal and those that aren't tend to conserve energy by seeking the shade, where they are hard to spot, during the heat of the day.

The early explorers found abundant wildlife. During his overnight camp at Chambers Pillar in 1872, for example, Ernest Giles saw a thorny devil "a most peculiar lizard ... armed at all points and joints with stout spines" and discovered a possum asleep in a crevice of the pillar.

Possums are one of a number of creatures whose numbers have declined drastically since European settlement. The central Australian race of the brush-tailed possum is now endangered, as are the bilby and the rufous hare-wallaby. Other species, including the lesser bilby and the burrowing bettong, are now extinct, although the latter survives on three Western Australian islands.

Introduced animals are largely to blame – particularly rabbits, which compete for burrows and food – but changes in the fire regime since Aborigines began leaving their traditional lands earlier this century have also had a major impact. When they managed the land by patch-burning to create a mosaic of areas of older vegetation, which animals relied on for shelter, and vigorous new growth, which provided food, destructive large-scale fires were rare because the burnt areas acted as firebreaks.

Pastoralists and park managers now recognise the importance of fire, but often lack the resources to implement widespread and effective burning programs.

Mammals

Three of the larger kangaroo species live in the Red Centre and the biggest – the red kangaroo, which favours mulga woodlands and tree-dotted plains that offer some shade – is particularly well adapted to the arid conditions. It is highly mobile and in times of prolonged drought the male becomes sterile and the female's reproductive system shuts down.

Like the other two species – the euro, or common wallaroo, which is found in the rocky hill-country, and the black-footed rock wallaby – it shelters from the sun during the day, emerging to feed on grasses at dusk. The black-footed rock wallaby is now considered vulnerable, but they can often be seen at Simpsons Gap and elsewhere

One of the most remarkable of the more frequently seen mammals is the black-footed rock wallaby (above), whose agility makes it superbly adapted to its rocky habitat high in the gorges of the MacDonnell Ranges.

Dingoes (opposite) are opportunistic scavengers and although mainly nocturnal, they'll often patrol popular tourist spots during the day. Their coats range from a pale sandy colour to reddish brown and, occasionally, black.

Although believed common in the desert ranges, the resourceful fat-tailed antechinus (top) is rarely spotted because it feeds mainly at night. Insects make up the bulk of its diet and it stores surplus fat in its tail, which is nearly as long as its mouse-sized body. Like emus (above), which roam throughout the Red Centre, they were much prized as food by Aborigines.

in the ranges where their agility helps them to keep one jump ahead of predatory dingoes, foxes and wedge-tailed eagles.

The dingo is probably the mammal that visitors are most likely to see. They'll often patrol popular camping spots in the evening looking for scraps to supplement their staple diet of rabbits and rodents.

All visitors are likely to see of the myriad smaller desert mammals are their delicate tracks in the sand. Some, like the spinifex hopping-mouse and the sandy inland mouse, shelter in burrows up to a metre underground, emerging at night to feed on seeds, grass and roots. Others, like the carnivorous stripe-faced dunnart and the fat-tailed antechinus, shelter under rocks and logs and feed at night on insects, storing surplus fat in their tails for when food is short. All can survive without drinking water.

Birds

Birds are by far the most conspicuous creatures in the Red Centre and visitors who've driven there along the Stuart Highway will frequently spot Australia's largest bird of prey, the wedge-tailed eagle, feasting on dead animals by the roadside. With a wingspan that can reach 2 m and an ability to fly at up to 95 km/h, wedgies can be distinguished from the region's many other birds of prey – such as kites and the much smaller brown falcon – by their feathered legs and distinctive wedge-shaped tail.

The most colourful – and noisy – birds are the parrots and cockatoos. Anyone who has camped near river red gums will know they're a favourite roost for huge flocks of galahs and little corellas, and smaller numbers of Port Lincoln ringnecks, who fly off noisily at sunrise to feed on flowers, fruit and seeds. Pink cockatoos, commonly called Major Mitchells, and flocks of red-tailed black cockatoos and budgerigars can often be seen feeding on seeds and grasses by the roadside.

Woodland birds, like the mistletoe bird with its distinctive scarlet breast, the grey shrike thrush and pied butcherbird – both of which have clear, flute-like songs – the spotted bowerbird and the honeyeaters, are easiest to see in the cool of the morning or late afternoon.

Spinifex pigeons, with their barred chestnut markings, red eye patch and distinctive head plumes, are among the most conspicuous birds of the rocky hillsides. Others include the crested pigeon, the willie wagtail, the zebra finch and the painted firetail. The latter two species can often be seen congregating at waterholes towards dusk.

Perhaps the most unexpected of the Red Centre's 200-plus species of birds are the waterbirds. Some, like pelicans and black swans, use the region's waterholes as stopovers on their migratory flights, but many – including black ducks, grey teal, herons and cormorants – will stay as long as water is plentiful.

Reptiles and frogs

Reptiles are superbly adapted to life in the arid zone and the most spectacular of these is the perentie, Australia's largest lizard at up to 2 m long. Like many other reptiles it

The permanent waterholes that dot the Red Centre are a lifeline to many creatures, including nomadic birds like pelicans (left) and black swans (top) that use them as stopover points as they criss-cross Australia. More permanent winged residents include galahs (above), which often congregate in noisy groups as they feed on seeds or prepare to roost in the trees lining watercourses.

The rough-skinned central netted dragon (above) regulates its body temperature by dividing its time between sunny perches and shallow burrows. The 2 m long perentie (above right), our largest lizard, can appear intimidating as it flicks its tongue from side to side, but will usually ignore humans if they give it a wide berth. Donkeys (right) are among a number of feral animals introduced by Europeans.

hibernates during winter, but at other times it can be seen patrolling rocky hillsides for the small mammals and birds that make up the bulk of its diet.

Among the hundreds of other reptile species are the skinks, geckos, dragon lizards and sand goannas. Each is superbly camouflaged for its desert environment and it takes patience and a keen eye to spot many of them.

Snakes – including four species of python, the venomous mulga, or king brown, which grows to 2.5 m long, and western brown snakes – are also common throughout the Red Centre. They usually hunt by day during winter and should be given a wide berth.

Visitors will be lucky to see any of the frogs found in the region. Many burrow up to a metre below the sand for long periods, encased in a water-holding shroud of their own skin. They're awakened by heavy rain percolating through the sand and emerge briefly to breed and feed before returning to their subterranean world.

Insects and spiders

It's often easier to spot insects' homes than it is to spot the tiny creatures themselves. What sometimes look like misshapen fruits hanging from trees are the silken homes of bag shelter moth caterpillars, known locally as itchy grubs. They rest in the protective bags during the day and emerge in processions at night to feed on leaves.

Mulga ants build distinctive, raised, halo-like nests up to 20 cm in diameter that they cover with mulga leaves to guard against flash flooding. Smaller holes in the ground, up to 10 cm in diameter, are the homes of barking spiders, which emerge at night to hunt for small insects.

Two insects that are a delicacy for the desert Aborigines are witchetty grubs, which feed on the roots of acacias and whose discarded pupal shells can sometimes be seen on the ground, and honey ants, which burrow 2 m deep in mulga woodland.

Fish

Probably the most surprising creatures in the arid zone are the fish that cling precariously to life in the permanent waterholes. In spring and summer the females lay as many as 100,000 eggs, but few of the hatchlings survive to adulthood because they're eaten by other aquatic creatures. The largest of the Red Centre's 11 species of fish is the bony bream, which grows to 30 cm, and can often be spotted in Ormiston Gorge.

Introduced animals

Australia is home to an estimated 60,000 feral camels, the descendants of creatures imported from 1840 to open up the continent. Groups can often be seen roaming the dune country to the south of Alice Springs.

Other prolific introduced animals in the region are feral horses, known as brumbies, rabbits, foxes and cats. Brumbies and rabbits are both blamed for soil erosion and overgrazing, and rabbits also compete with native animals for shelter. Although cats and foxes eat rabbits, they also kill native birds, mammals and reptiles.

PIP SMITH

The distinctive nests (top) of the mulga ant are a common sight throughout the Red Centre. So too are camels (above), descendants of the estimated 10,000 beasts imported last century to open up the interior of Australia. When improved rail and road transport made them redundant in the 1930s, they were allowed to run wild and their numbers have risen to at least 60,000.

Plants and their communities

The Green Centre

Oasis in the desert. Although arid, the Red Centre landscape is full of surprises – like this permanent water-hole on the Finke River to the south of Alice Springs. The waterhole, in the heart of a cattle station, is fringed by river red gums that can grow up to 30 m tall and is home to a profusion of birdlife, including pelicans. The abundance and diversity of vegetation in the region surprises and delights most visitors – and is at its most striking after rain. But even in times of drought

Most first-time visitors to the Red Centre are amazed by the profusion and diversity of plant life. It's particularly startling for those arriving by air because they'll have looked down on a predominantly red landscape but be greeted at ground level by one that – except after bushfires or prolonged droughts – is almost as green as it is red.

The abundant greenery is all the more remarkable because the Red Centre has an average annual rainfall of less than 250 mm, the threshold usually regarded as defining a desert, and its most arid area, the Simpson Desert, receives an average of just 100 mm a year. Both figures mask the erratic and unpredictable nature of the rain; several years of above-average rainfall may be followed by drought.

Coupled with erratic rainfall and low humidity are extreme temperatures, ranging from a low of –7°C on winter nights, when there are frequent frosts, to summer highs that regularly exceed 40°C.

Plants have developed ingenious ways of coping with the extreme heat and lack of water. Small, spiny leaves with waxy surfaces minimise moisture loss, and extensive root systems – often twice as big as the visible part of the plant – gather whatever water is available.

In times of drought the river red gums lining many sandy watercourses shed limbs to conserve moisture, and some plants, especially annuals like daisies and pussy tails, remain in the ground as dormant seeds until rain – and even then they hedge their bets by not all germinating at once.

Many plants are adept at coping with fire. Desert eucalypts, for example, have thick, flaky bark that insulates them, or smooth, white bark that reflects the heat. Those shrubs that don't survive fire well, like mulga, produce large quantities of seeds that need fire to start the germination process, thus ensuring their survival.

There are three main plant communities in the region – the desert ranges, the dunes and sand plains, and the mulga woodlands. There are also two smaller, but very distinctive, communities – the gibber and saltbush plains and the salt lakes.

The Desert Ranges

The most striking features of the Red Centre are the dramatic mountain ranges that rise up to 1000 m from the otherwise flat plain. At first glance their rugged spinifex and shrub-covered flanks and craggy summits appear inhospitable, but nestled within them are a wealth of different plant communities.

the vegetation is impressive because plants have adapted survival strategies that help them cope with the extreme heat and lack of water. Some, like the river red gum, shed limbs to help them conserve water, while others, like mulga, funnel any moisture down their trunks into the ground. Leaves and roots play their part too, ensuring that far from being barren, the arid zone is every bit as interesting botanically as the better-watered parts of our continent.

The parallel ranges of the West MacDonnells National Park, for example, are home to almost 600 plant species, 75 of them rare. The Chewings Range is particularly important, containing six species found nowhere else in the world and 18 that date from a much wetter era.

The best-known haven for relict plants is Palm Valley, where 12,500 red cabbage palms rise from the valley floor. Their shallow roots tap into the water that filters into the valley from the porous sandstone hills.

Like the ancient MacDonnell Ranges cycads that are found nearby, as well as in several sheltered gorges in the MacDonnell Ranges and Watarrka National Park, they are remnants of patches of rainforest that were widespread some 65 million years ago.

The dominant ground-cover plants on the lower slopes of the ranges are the spiny grasses commonly known as spinifex. There are more than 20 species, generally divided into soft and hard varieties according to their degree of prickliness, and they are interspersed with other grasses and herbs like the common daisy. These slopes are also dotted with ruby saltbush, mulga, mallee, witchetty bushes, fragrant yellow-flowered

Some of the most surprising plants in the Red Centre are the MacDonnell Ranges cycads (above) that flourish in a handful of sheltered gorges and are relicts from a much wetter age. More typical are the river red gums that flourish at Serpentine Gorge (right) and the ghost gums (opposite) that cling tenaciously to life in some of the unlikeliest locations.

The smooth bark of river red gums is often patterned with patches of red and grey (above). Regarded by many people as the symbol of inland Australia, it is probably our most widely distributed eucalypt, being found along watercourses in every mainland State. Less familiar in the Red Centre is the grass tree (right). A mass of stiff leaves up to 1 m long sprout from the top of its fire-blackened trunk, forming a skirt around it as they wither. Its spear-like flower spike may be up to 3 m long.

cassias and the taller white cypress pines, whitewoods and bloodwoods. Near water-courses, tea-trees, river red gums and ghost gums are common. The latter also grow in many rocky gullies along with white cypress pines.

The dunes and sand plains

The dominant Red Centre landscape, particularly in the area to the south of Alice Springs, is the dune and sand plain country. Many of the dunes are little more than bumps, but on the eastern boundary of the area covered by this book they rise to 15 m.

Here spinifex predominates. With roots that may reach down more than 3 m and spiky leaves that curl inwards at the first sign of drought, the plant is superbly adapted to its desert environment. All species are highly flammable but they recover well from fire, quickly recolonising the landscape.

The trees that dot the spinifex country – like the grevilleas with their bright yellow flowers and the stately desert oaks – survive fire well because they are protected by thick bark.

The dunes and sand plains are some of the best places to appreciate the delicate flowers that carpet the sand with a mantle of yellows, whites, pinks and purples after

Spinifex (above and left) is the popular name for more than 20 species of spiny tussock grasses that flourish throughout the Red Centre. With long roots and spiky leaves that curl inwards to conserve moisture, they are superbly adapted to their arid environment. Some species form huge prickly rings, up to 10 m in diameter, as the centre of the hummock dies and new growth appears around the edges. Softer species, like the graceful feathertop spinifex, resemble a golden cereal crop as their seed heads sway in the breeze.

Majestic desert oaks (above) begin life as gawky youngsters, like the two in the distance, then branch out as they grow to a height of 10 m. Their cones, which are the size of a Kiwi fruit, are a favourite food of termites. Like white cypress pines (above right), which are found in rockier country, they have thick bark and long, narrow, needle-like leaves that minimise moisture loss. Although cypresses can grow to 20 m, those in the arid zone rarely grow to more than 10 m and in exposed positions often take on a gnarled appearance.

rain. These include yellow-top daisies, poached egg daisies, papery everlastings and fluffy pussy tails. Bright pink desert fringe-myrtles and the purple-flowered, fleshy-leaved parakeelyas, which are also common in these communities, are less dependent upon rain.

Mulga woodlands

Mulga is by far the most common shrub in the Red Centre, its dense grey-green leaves and yellow flowers dominating the landscape to the north of Alice Springs. It's also found on the lower slopes of some of the ranges, in the depressions between dunes and even on the top of Uluru.

Mulga is well adapted to making the most of the little rain it receives. Its tiny leaves point upwards and rain is channelled along the branches, then down the trunk of the tree directly to the root zone. This clever funnelling means water penetrates the ground more deeply and is less likely to evaporate than if it were spread more widely and thinly.

Even so, some trees die during prolonged drought and many more fall victim to fire – although fire helps to germinate the trees' abundant seeds. Mistletoe, a parasitic plant with red flowers and sticky berries that are spread by mistletoe birds, is often found

Minnie daisies (left) are among many wild flowers that add vibrant splashes of colour to the desert landscape after winter rain. Their seeds may have lain dormant in the ground for months or even years, but they'll germinate within a few days of rain and quickly flower. Paddymelons (below) are introduced vines that flourish in disturbed ground, particularly along the sides of roads, throughout the region. The seeds of the ripe melons are a favourite food of the larger parrots but are too bitter for humans. Territorians often use them as substitute balls in a variety of sports and one caravan park in the region hosts a regular Saturday evening paddymelon-bowls tournament.

growing on mulga. Common flowers in mulga woodlands include short-lived sunrays, everlastings and pussy tails, all of which germinate after rain.

Gibber and saltbush plains

The gibber and saltbush plains to the south-east of Alice Springs appear inhospitable, but their soil is rich in nutrients. Here the dominant plants are some of the 100 species of saltbush and bluebush, which have grey-green or blue-green leaves that help reflect the sunlight to keep the plants cooler. Their thick, hairy leaves prevent them from drying out in strong winds.

Trees are rare but a stand of spindly looking waddy-wood trees thrives in gibber plains on the western fringe of the Simpson Desert (see Chapter 11).

Salt lakes

The shimmering salt lakes in the south-west of the region probably have the sparsest vegetation of any of the plant communities, with the most common plants being small, salt-tolerant samphires. Their salty, succulent leaves and stems can be cooked as a vegetable.

Alice Springs
A town like Alice

The hub of the Red Centre. In less than 50 years, Alice Springs has been transformed from a dusty frontier town to a thriving metropolis of more than 25,000 people. Much of its rapid growth can be attributed to tourism, because the town is the gateway to a region that attracts almost half a million visitors a year, but it's also an important regional and commercial centre.

If Nevil Shute was to revisit the town he immortalised in his classic novel *A Town Like Alice*, he probably wouldn't recognise the place. Yet the milk bar in Todd Street that obviously made quite an impression on him when he passed through in 1949 is still standing – albeit in a different guise.

Brendan Heenan remembers it well. His father Mick opened it in 1949 and as a young schoolboy, Brendan often helped out by making milkshakes. "It was pretty big and flash with lots of chrome and Laminex, which was all new after the war," he said fondly as we sat in a shady corner of the caravan park he owns in Alice Springs. "We opened seven days a week from 8 a.m. until 10 p.m. and it was the only place for socialising, apart from the pubs."

As well as selling ice-creams and milkshakes, it sold fruit and vegetables grown by Mick. Brendan doesn't recall the novelist's visit and he was at school in Adelaide in 1956 when film crews visited the milk bar – now a pharmacy – to shoot scenes for the movie based on the book. But he's in no doubt that the business Jean Paget dreamed of opening was modelled on it. "I want it to be a sort of ice-cream parlour with a few chromium-plated chairs and glass-topped tables," she noted. "I want to sell fruit there and fresh vegetables."

Today, tourists looking for the frontier town immortalised by Nevil Shute will be disappointed, for "the Alice", as it is affectionately known, has come of age and is now a modern town with shops and services to rival those of any of Australia's regional centres. But its colourful history comes alive in a number of its tourist attractions and it's well worth spending a few days there to soak up the unique atmosphere.

Alice Springs Telegraph Station
Sitting under a river red gum on the grassy bank overlooking Alice Spring, the small waterhole on the Todd River after which the town was named, it's not hard to imagine what life was like in 1872 for the first European settlers in the Red Centre.

They were Johannes Mueller, station master at the newly opened Telegraph Station beside the waterhole, 3 km north of the present town, and his five staff. They lived in simple stone barracks that also served as the telegraph office until more buildings were added – including the station master's residence and a post and telegraph office – and the telegraph station became a bustling community with its own cattle, vegetable gardens, forge and horse yards.

This early evening view of "the Alice", taken from West Gap and looking north-east, highlights the town's road system, but two other important "highways" are also clearly visible. They're the usually dry Todd River, to the right, bordered by river red gums, and the railway line, which runs parallel to the dual-carriageway Stuart Highway in the centre.

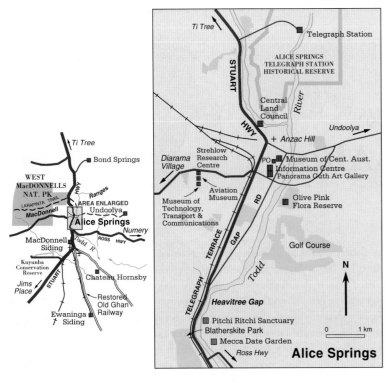

Alice Springs

The buildings have been meticulously restored and displays chart the evolution of the station during its 60 years as a vital communications link and its next 30 years – when it was a home for children of mixed Aboriginal and European descent, then a camp for Aborigines working for the Army and, finally, an Aboriginal settlement.

Stuart Town Gaol

This spartan prison, built in 1908 from stone quarried near Heavitree Gap, is the oldest building in the town centre and was constructed by stonemason Jack Williams, who also built Adelaide House.

Until 1938, it housed petty criminals serving sentences ranging from a few days to a few months for crimes that included cattle duffing, drunkenness, being idle and disorderly, and trying to pass valueless cheques. The last two inmates were jailed for travelling on the Ghan without tickets.

Adelaide House

This simple stone building, which opened in 1926 as the Red Centre's first hospital, was designed by the Rev. John Flynn – who later founded the RFDS – and featured a revolutionary cooling system in which wet sacking filtered out dust and cooled the air as it circulated.

The nurses who staffed it had to walk to the telegraph station and consult the nearest doctors at Darwin and Port Augusta by telegram, but it represented a major advance in medical services. Until 1915, when Sister Finlayson arrived in the town, anyone requiring medical attention had to make the arduous two-week buggy journey to Oodnadatta.

Adelaide House ended its role as a hospital in 1939 when a new hospital opened, but it served for many years as a hostel for pregnant bush women.

The Central Australian Aviation Museum

Aviation played a vital role in the growth and development of Alice Springs from the 1930s. This museum – next to the spot where the first plane, carrying writer and explorer Francis Birtles, landed in the town on 5 October 1921 – charts the impact of aviation.

That first aircraft was a far cry from the jets that now link Alice Springs with all the major cities in Australia – before it could take off, it had to wait for a camel train to deliver fuel. In 1935 Alice Springs became the overnight stop for the new two-day commercial flight from Adelaide to Darwin and four years later, aviation pioneer Eddie Connellan began the first mail run.

Among the exhibits are a RFDS aircraft that made many live-saving flights, and other early aircraft. Beside the museum is a display housing the remains of the ill-fated *Kookaburra*, lost in the Tanami Desert in 1929 and recovered in 1978 by Dick Smith.

Be sure to wander into the neighbouring hangar, which houses the Museum of

The tragic story of the KOOKABURRA, the little monoplane forced down with engine trouble in the Tanami Desert in 1929, is brought alive in a special display (left) at the Central Australian Aviation Museum. Although pilot Keith Anderson and engineer Bobby Hitchcock managed to repair the engine, they died of thirst trying to hack down the thick scrub that surrounded their plane. Other displays in the museum include the JOHN FLYNN (below), a RFDS plane named after the founder of the medical service that still provides a mantle of safety for the people of the outback. The aircraft was a far cry from the transport used by the early sisters (opposite) at Adelaide House, the region's first hospital, which Flynn established in 1926.

MIKE GILLAM

A walk through Alice Springs Desert Park (above), 6 km west of Alice Springs, gives visitors the chance to understand the rich biological diversity of the Red Centre. This 1300 ha park, a tourist and research facility for rare and endangered native plants and animals, features examples of each of the region's main landscapes.

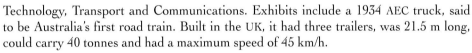

Technology, Transport and Communications. Exhibits include a 1934 AEC truck, said to be Australia's first road train. Built in the UK, it had three trailers, was 21.5 m long, could carry 40 tonnes and had a maximum speed of 45 km/h.

The Strehlow Research Centre

Most museums proudly display their riches, but visitors to the Strehlow Research Centre won't see the bulk of the unique collection of artefacts, photographs, film and documents amassed by Professor Ted Strehlow during a lifetime spent living and working with the Arrernte people.

The striking $3 million centre, which features the biggest rammed-earth wall in the Southern Hemisphere, was built to celebrate the work of Ted Strehlow, who grew up at Hermannsburg and became an authority on Arrernte culture and traditions. But much of his vast collection, one of the most significant and comprehensive in Australia, is behind lock and key in eight security zones because of its sensitive nature.

David Hugo, the centre's research director, told me that only he and one other staff member have seen the entire collection, which includes 1200 sacred objects, 25 hours of 16 mm film, 6000 photos, 25 hours of sound recordings, letters, diaries and maps. "The Aborigines stipulated that only people with a serious interest were to have access," he said.

Although visitors don't see most of the collection, they can still gain a valuable insight into Arrernte life and learn about the work of Ted Strehlow, who was invited to document Arrernte culture in the 1930s by elders who feared that otherwise it could die with them.

The hunter-gatherer lifestyle of the Red Centre's Aborigines (above left) was fast disappearing by 1933, when Ted Strehlow was invited by a group of elders to document aspects of their culture. His father, Pastor Carl Strehlow, ran the Lutheran mission at Hermannsburg for 28 years and from the age of two, until he went to school in Adelaide at the age of 14, the Arrernte children were his only playmates. Later, elders treated him like a ceremonial chief at meetings (above) and ceremonies – entrusting him with sacred objects and artefacts that now form the core of a unique collection housed at the Strehlow Research Centre, shared with the Museum of Central Australia, in Alice Springs. Visitors to the imposing centre gain a fascinating insight into Arrernte life and the work of a remarkable Australian through integrated displays at the Museum of Central Australia.

Within the Museum of Central Australia, photographic displays detail the life and work of Ted Strehlow, whose father, Pastor Carl Strehlow, ran the Lutheran mission at Hermannsburg for 28 years.

Panorama Guth

"I paint non-stop," Henk Guth told me as we surveyed his masterpiece, *Panorama Guth*, which stands 6 m high, is 60 m in circumference and depicts every major landmark in the Red Centre.

And I could believe him. Henk spent just six months painting his panorama in 1975, with assistance from fellow Dutch artist Frits Pieters. The 33 panels of Irish linen were covered, in situ, with an estimated 680 kg of oil paints and are now admired annually by about 90,000 visitors, who climb a spiral staircase to a central platform that simulates a lookout.

It's the only panorama in Australia and one of 33 in the world. "I would like to make a panorama of the whole country," Henk told me. "But the cost would be tremendous and I don't think it will ever be done."

As well as showcasing the panorama, Henk's gallery in Hartley Street displays many of his original oils, which are for sale. Visitors can also see watercolours from the Hermannsburg school, including some by Albert Namatjira, and Aboriginal artefacts.

The Old Ghan

It was hard to tell who was enjoying themselves most as the old Ghan train halted at Mt Ertiva Siding – train driver Laurie Nicholls or the 60 passengers he was taking to

The name of fortune-seeker Harold Bell Lasseter, who claimed he found a fabulously rich gold reef in the Petermann Ranges in 1897, is immortalised in the casino (above) in Alice Springs where up to 350,000 people a year try their luck at games like roulette, blackjack and two-up, and on the slot machines. Lasseters Hotel Casino opened in 1981, 50 years after Lasseter died trying to relocate his lost reef, and caters for all types of gamblers from those wanting a small flutter on the slot machines – where the biggest-ever payout was $462,000 – to those playing for high stakes.

Visitors who don't have time to take in all the Red Centre's many natural attractions can see what they're missing by climbing the spiral staircase at Panorama Guth (right). The creation of Dutch-born artist Henk Guth (opposite), the panorama depicts – in oils on linen – every major landmark in the region and is the only one in Australia.

dinner in the bush. Laurie had driven the train every day for the past 10 weeks but from the broad smile that lit his round face, I'd never have guessed. "I always get a thrill from it," he admitted as he clambered down from the engine. "Steam is a living, breathing thing."

Laurie has the old Ghan in his blood. His father worked in almost every fettlers' camp as a cook for the railway gangs that maintained the line after it was built in 1929, and Laurie fired up the train on its run from Quorn in South Australia to Alice Springs for four years from 1947. Later he drove the new standard-gauge Ghan, which took over in 1980. "It doesn't have the same flavour," he said. "On the old Ghan you'd wake up in the morning covered in soot."

Now 69 and retired, he spends about four months of the year driving the old Ghan – as a holiday. He told me he'd been up at 4 a.m. to shovel coal into the engine's firebox and that he wouldn't get to bed until after midnight, but he wasn't complaining. He's one of about 10 volunteers actively involved in the Ghan Preservation Society, which was formed in 1981 and started operating tourist services in 1988.

During a camp-oven dinner of roast beef and roast vegies under the stars in a patch of scrub next to the track, I discovered that Laurie wasn't the only train buff travelling on the Ghan that night. For the AG members from Queensland's Fraser Island sharing my table, it was one of four trains they'd be travelling on during their annual holidays. An American group said the trip was the highlight of their time in the Red Centre – both for the train ride and the outdoor dinner, which gave them their first opportunity to admire the star-studded sky.

Earlier, as the train pulled out of Stuart Station – built in 1988 with a Bicentennial grant – on its leisurely 9.5 km journey, I sat with society secretary Allen Orr in comfortable armchairs in the elegant bar car, the only carriage that was part of the original Ghan rolling stock. "We have a lot more stock that can be refurbished," he said. "But apart from two paid staff who maintain the track and the engines, we rely on volunteers and our only funding is from donations, charters and ticket and souvenir sales."

Chateau Hornsby

As soon as the last strains of *Auld Lang Syne* have heralded the New Year, Denis Hornsby sends his party guests into his vineyards to pick grapes. Four months later, the fruits of their labour produce what Denis claims is the world's first vintage of the year: his Chateau Hornsby early-bottled shiraz.

Denis established his 2.8 ha of vineyards in 1974 on former citrus orchards on the south-east outskirts of Alice Springs and six years later gave up a career as a pharmacist to concentrate on making the Red Centre's only wines. He now produces about 18,000 bottles a year, most of which are sold to tourists at the cellar door.

His five varieties of grapes – shiraz, cabernet sauvignon, Rhine riesling, semillon and chardonnay – ripen before crops in Australia's traditional wine areas further south and the first are picked a few minutes into the New Year. "We pick about half a tonne

The golden age of steam. Oilcan in hand, railway buff Laurie Nicholls (above) prepares to service the 1951-vintage Western Australian steam engine that heads the old Ghan train on regular tourist runs from Stuart Station, on the outskirts of Alice Springs. A third-generation railwayman, Laurie spends up to four months a year of his retirement driving the train for the Ghan Preservation Society, a voluntary group that fought successfully to preserve a section of the old narrow-gauge railway when it was closed in 1980. Later (opposite above), with a puff of smoke, Laurie pulls away from Stuart for the leisurely 9.5 km run to Mt Ertiva Siding. Passengers travel in comfortable carriages, including the plush bar car that was part of the original Ghan rolling stock, or on the popular flat-top wagon (opposite below). Many are train enthusiasts and Laurie loves telling them stories about his days as a fireman on the old Ghan line and a driver on its standard-gauge successor.

of grapes by the light of car headlights and floodlights," he told me as we sampled the 1994 vintage on a shady terrace overlooking the vines. "It's a token gesture – we get the rest later when we can see what we're doing – but it makes it the first vintage in the world. We don't necessarily get it into the bottle first, but we certainly pick and make it before anyone else."

For the past five years, the winery has also been the venue for the thrice-weekly *Ted Egan Show*, which attracts audiences of up to 100. Ted's affection for the people he's met in his 40 years in the Territory is infectious and during his two-hour show, which he describes as a mix of "singing songs and telling lies", he introduces many of them in compositions sung to the accompaniment of the "Fosterphone", the empty beer carton he uses to tap out a beat.

Ted settled in Alice Springs in 1977 when he gave up his 27-year career in the public service, much of it spent in remote bush postings. Between his six-month seasons at Chateau Hornsby and touring engagements in Australia and overseas, he's working for his master's degree in history, writing the second part of his autobiography, compiling material for his ambitious *Faces of Australia* recording project and, with his wife Nerys Evans, writing the script for a major film that shares its theme with his song, *The Drover's Boy*. The film, based on the story of an Aboriginal girl who survived the Coniston massacre in 1928 and was made to work "disguised" as a boy by day and be a "wife" by night, is typical of Ted's meticulous research and his great compassion towards Aborigines.

He was awarded the Order of Australia in 1993 for his services to Aboriginal communities and his contribution to Australia's cultural heritage, and spends much of his

Entertainer Ted Egan (right) taps out a beat on his "Fosterphone", the instrument he invented because, in his own words: "I'm a compulsive finger tapper and I can't play anything else." Three evenings a week every winter, he takes over the intimate cellar-bar at the Chateau Hornsby winery, on the outskirts of Alice Springs, to entertain audiences with outback songs and stories based on the people he's met in 40 years in the Territory. Winemaker Denis Hornsby (opposite left), pictured sampling his early-picked shiraz, grows five varieties of grapes (opposite right) and begins harvesting them a few minutes into each new year to produce what he claims is the world's first vintage of the year. Grape pickers (above) work in the relative cool of early morning to beat midsummer temperatures that may soar to about 50°C in the shade.

time acting as an unofficial ambassador for the Territory. It's a role he's ideally suited to as he readily admits he loves both Darwin and Alice Springs. "Alice is an exciting place," he told me. "We get together with dynamic people on a regular basis."

Bond Springs

Bond Springs is 18 km north of the Alice and, at 1515 sq. km, is a small cattle station by Territory standards. But what it lacks in size it makes up for in hospitality. Since 1990 Jan Heaslip has invited visitors to join in its day-to-day life as a working station and soak up the historic atmosphere in some of Australia's oldest and best-preserved station buildings, dating back to the 1870s.

Jan and her husband Grant bought the Bond Springs lease – once owned by cattle king Sir Sidney Kidman – in 1964 to breed poll Herefords. They now have about 3000 breeders but they've diversified with a wholesale meat company and tourism.

"I started with people coming out for lunch and a tour, and they were constantly asking if we had accommodation," Jan told me over coffee in her large, homely, timber kitchen. "Rural tourism is becoming quite important and I decided I should be in it." To accommodate guests, many of whom come from North America and Europe, she converted a spare bedroom in the homestead to a guest room and renovated a three-bedroom cottage in the grounds.

Jan's love for Bond Springs and its history was obvious as she took me on a tour that began in the old timber homestead, built of poles in the 1870s. The first family to live in this one-roomed building with its rough stone floor and corrugated-iron roof spent more than a year getting there from Adelaide.

The annual Camel Cup (opposite), usually held in May, continues a tradition begun in the late 1800s by Afghan cameleers, who often amused themselves with impromptu races. The beasts responsible for opening up much of the Red Centre are erratic racers, but that all adds to the fun for their riders and the thousands of spectators who pack into Blatherskite Park. The big winners are the local Lions clubs, which raise thousands of dollars for community projects.

The historic and meticulously maintained outbuildings at Bond Springs station give visitors a fascinating glimpse into the past. The quaint schoolroom (below), built of the plentiful local stone, houses the old wireless used by owners Jan and Grant Heaslip's four children for their lessons before they went to boarding school in Adelaide. The leather room (below right) is crammed with harness, saddles, camel packsaddles and a leather water-carrier believed to have been brought to Australia by an Afghan cameleer. Each item has been lovingly preserved by the Heaslips, who enjoy sharing station life with a growing number of visitors.

TONY STANTON

TONY STANTON

One of the most novel and exciting ways to see the Red Centre is from the gondola of a hot-air balloon (above) as it drifts up to 700 m above the southern flanks of the MacDonnell Ranges. The cool, crisp mornings that are a feature of the Alice Springs climate are perfect for ballooning and 30- and 60-minute flights are followed by champagne breakfasts in the bush. Camel trekking (right) is another special way to build up an appetite and rides along the sandy Todd River bed end with breakfast or dinner. Camels leave from the Mecca Date Garden, a shady oasis offering fresh dates from trees originally planted by Afghan cameleers and moved to their present site in the 1960s.

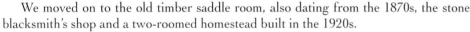

We moved on to the old timber saddle room, also dating from the 1870s, the stone blacksmith's shop and a two-roomed homestead built in the 1920s.

Nearby were the old stone schoolroom, where Jan's four children did their School of the Air lessons, and the leather room, perhaps the most atmospheric building on the property. It was crammed full of branding irons, old harness, saddles and camel pack-saddles – all lovingly polished and giving off the rich aroma of well-worn but cared-for leather.

I realised that what was special about Bond Springs was that nothing had been thrown away; rather it had been restored, with generous applications of elbow grease, and displayed to evoke a sense of history.

Other attractions

Among a host of other attractions in and around the Alice are hot-air balloon flights to watch the sunrise wash over the ridges of the east MacDonnell Ranges, camel rides along the bed of the Todd River, the Olive Pink Flora Reserve, where arid zone plants are displayed, the Pitchi Richi Sanctuary, which gives a taste of Aboriginal culture, and the Museum of Central Australia, where the fascinating geology and fossil life of the region are brought to life. The School of the Air and RFDS (see Chapter 4) are also well worth a visit.

If you're in town in early May, don't miss the annual Camel Cup at Blatherskite Park. Six months later, the Todd River is the venue for an even crazier event, the Henley-on-Todd Regatta, when contestants race their bottomless "boats" along its dry bed using leg power.

On the rare occasions when the Todd River flows (above), Alice Springs locals waste no time in dusting off their canoes and kayaks and taking to the water. Paddling down the knee-deep river towards the town centre are, from left, Brooke O'Connor, Daniel Mostran, Peter Mostran and Scott Murdoch. The river usually flows for no more than a few days each year, but outback humour and ingenuity have made the annual Henley-on-Todd Regatta, held on the last Saturday in September or the first in October, the biggest event on the Alice Springs calendar. Started by local Rotarians in 1961 to raise funds for community projects like fitting smoke detectors in the homes of the elderly, it has raised more than $600,000 and is so popular that mini regattas – like this one at Simpsons Gap (above left) staged for a tourist convention – can be organised for large groups of visitors. Contestants who run into trouble in their bottomless boats need have no fear of drowning in the sand – lifesavers are always on hand to rescue them!

West of Alice Springs
A walk on the wild side

Namatjira country. Dominated by a series of parallel ridges and extending 160 km west from Alice Springs, the timeless grandeur of the west MacDonnell landscape was immortalised in vibrant watercolours by Albert Namatjira, one of our best-known Aboriginal artists. The recently created West MacDonnells National Park protects 210,000 ha of the most spectacular country – including the highest peaks west of the Great Dividing Range – and a number of important refuges for rare and

The late afternoon sun was painting the high rock walls a burnished copper as I entered the shady confines of narrow, boulder-strewn Inarlanga Pass, 101 km west of Alice Springs. I picked my way along the deep gully over a jumble of pinky-grey quartzite rocks, relishing the coolness after more than an hour of traversing the exposed, 1000 m high ridge tops of the Heavitree Range in temperatures nudging 30°C.

I slipped off my pack and perched on a smooth, flood-polished boulder to take in the extraordinary beauty of my surroundings. Lush MacDonnell Ranges cycads sprouted from crevices on the floor of the chasm, their palm-like leaves arching gracefully for more than a metre from trunk to tip. Dotted among them were ghost gums, delicate ferns and grasses and, higher up the glistening orange quartzite walls, a profusion of small shrubs. I'd discovered one of the many oases nestled within the ancient mountain ranges of the West MacDonnells National Park.

What was so special about this one was that I had it to myself. The long east–west ridges of the west MacDonnell Ranges and the parallel Heavitree and Chewings ranges are punctuated by numerous deep gashes – carved by ancient water-

courses more than 300 million years ago – where relict plants from wetter eras flourish. Some, like Simpsons Gap, 15 km west of Alice, are visited by up to 150,000 people every year, but others are probably seen by fewer than 1000 people a year.

Inarlanga Pass falls into the latter category. I'd walked to it from Serpentine Chalet bush camp along a 2.5 km stretch of the Larapinta Trail, a world-class walking track that, when completed in 1998 at a cost of more than $1 million, will extend 220 km west from Alice Springs to Mount Razorback across some of the Red Centre's finest scenery.

The only other tracks I'd seen as I followed the trail along rocky ridges and through occasional pockets of mulga woodland had been those of dingoes. The views had been breathtaking. Pale green spinifex gave the rugged hills a deceptively soft grassy appearance as they stretched to the western horizon, where they met the hazy blue-grey outlines of some of the highest peaks west of the Great Dividing Range.

Apart from the narrow rocky path at my feet, there was no sign of human impact. I felt I was experiencing the ancient landscape exactly as the first Aborigines had, some 30,000 years earlier.

relict plants and endangered wildlife. Although it attracts some 180,000 visitors a year, many rush through in a day and venture only short distances from their vehicles. Anyone with time to spare can seek out numerous special places a little more off the beaten track and enjoy some of the best wilderness experiences the Red Centre has to offer. These include the 220 km Larapinta Trail, a world-class track that, when completed in 1998, will offer experienced bushwalkers the ultimate view of the park.

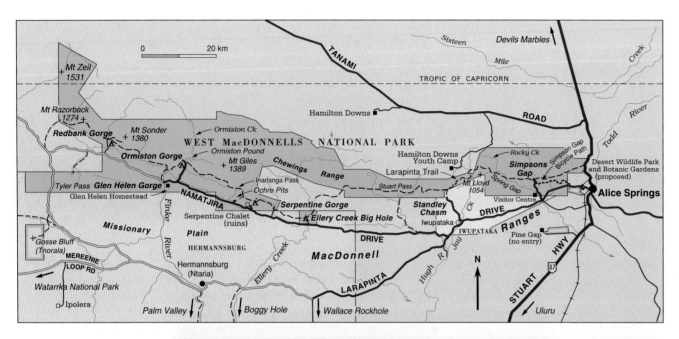

The flood-polished quartzite walls of tranquil Inarlanga Pass gleam in the late afternoon sun and walkers who pick their way over the jumble of rocks will find ancient cycads and delicate ferns and grasses sprouting from the many crevices. The shady chasm is one of numerous delightful rest stops for bushwalkers on the Larapinta Trail, but day walkers can easily reach it from the Serpentine Chalet bush camp or the ochre pits.

TONY STANTON

Ranger Chris Howard (above) leads walkers through mulga shrub-land near Arenge Bluff on one of the twice-monthly ranger-guided winter bushwalks along the easternmost sections of the Larapinta Trail. The two-day treks, taken over by a tour company in 1995, take much of the hard work out of bushwalking because camping equipment, food and water are transported to the overnight camp by a support vehicle. Unencumbered by heavy packs, walkers have plenty of time to take in the extraordinary beauty of their surroundings and to learn about points of interest on the way. In a section of track dubbed the bush supermarket (above left) because of its abundant fruits and seeds, ranger Andrea Hope, fourth from left, explains how Aborigines ground the seeds from a shrub known as dead finish to make flour.

The Larapinta Trail

"The track's not just for experienced bushwalkers," Gary Passmore said as he led me along a 24 km stretch of the Larapinta Trail between Jay Creek and Simpsons Gap, where he is ranger-in-charge. "The two easternmost sections are well defined so young families and just about anyone can walk them. In fact, I've taken people aged from six to 86 on ranger-guided walks."

My first foray onto the track to Inarlanga Pass a month earlier had highlighted its versatility. Bushwalkers can venture onto it for a couple of hours to find a tranquil picnic spot, for a couple of days to enjoy the delights of a remote bush camp, or for a couple of weeks to walk its length and have a complete wilderness experience.

One of the best ways for visitors with limited time to sample the trail is to join one of the two-day ranger-guided walks held each winter – as I discovered when I set off from Jay Creek with Gary and five other bushwalkers, all from the eastern States. Gary probably knows the eastern part of the trail as well as anyone, because he helped plan it and oversee its construction – and his enthusiasm for the track and the scenery it traverses is infectious.

We'd not been walking long when he took us through what he called the bush

HOWARD WHELAN

Mountain majesty. Experienced bushwalkers who follow the rugged track from Ormiston Gorge to the summit of 1389 m Mt Giles can awake to this magnificent view (above) across Ormiston Gorge to Mt Sonder, perhaps the most hauntingly beautiful of the Red Centre peaks.

Mt Giles dominates the Chewings Range (opposite), seen here bathed in sunlight while a gathering storm darkens the parallel ridge of the Heavitree Range.

supermarket, pointing out native plums and the bush equivalents of tomatoes, potatoes, bananas, oranges – all important tucker for generations of Aborigines. On my own, I would probably have walked straight past them, and the emu bush and caustic bush that were prized for their medicinal qualities.

While Gary was bringing the bush to life, seasonal ranger Chris Howard was driving a 4WD ute over rough service trails to our camp at Rocky Creek, a wide ribbon of sand lined by river red gums. By the time we arrived, in late afternoon, he'd unloaded our swags and tents, got the fire going and had the billy boiling and the dampers baking. All we had to do was shed our day packs, sit back and watch the setting sun ignite Arenge Bluff, a majestic red quartzite outcrop that stood sentinel over our camp.

The trail is divided into 13 sections, most designed as two-day walks, and there are vehicle access points and drinking water supplies at approximately two-day intervals. Although the entire track has been marked and several groups have already walked its length, only four sections (1, 2, 3 and 8) covering 87 km were officially open at the time of writing and another two (10 and 12), covering 12.5 and 16 km respectively, were due for completion by winter 1995. (The numbers rise from east to west.)

The Conservation Commission of the Northern Territory has produced a series of excellent leaflets giving background information on the geology, plants and animals of the trail, as well as detailed track notes, hints on planning a walk and details of water supplies and suggested camping areas. They're available from the commission and from park rangers. Because much of the trail is in remote country, intending walkers should register with rangers and make sure they notify them of their safe return.

West MacDonnells National Park

The orange cliffs and purple and brown rocks that rise like ramparts from the parallel ridges of the west MacDonnell Ranges may seem strangely familiar to many first-time visitors. So too are the spinifex-covered rocky hillsides and the dramatic plant-filled gorges intersecting the rugged ranges that extend some 160 km west from Alice Springs.

It's hardly surprising. For these and many other features of the 210,000 ha "West Macs", as the park is affectionately known, were immortalised in vibrant watercolours by Aboriginal artist Albert Namatjira, who grew up at nearby Hermannsburg.

You don't have to venture far into the park to realise how accurately Namatjira captured the landscape. His bold use of colour may appear overstated when his paintings are viewed in a gallery, but thumb through a book of them during a visit to the park and you'll realise they're spot on.

The first of a number of planned "greater national parks" in the Red Centre, the West Macs evolved in 1992 from a string of seven small parks and nature reserves – most focusing on easily accessible waterholes framed by spectacular gorges. These gorges and the rugged landscape around them provide a habitat for nearly 600 plant species and a wealth of arid zone wildlife, including 167 bird, 23 mammal, 85 reptile, five frog and 10 fish species.

TONY STANTON

Curved rock at Inarlanga Pass (above) tells a complex story of faulting and folding over the ages. The layers of Heavitree quartzite were formed about 900 million years ago as sand washed into the vast inland sea that covered the area gradually hardened. Later, when the sea became shallower, layers of limestone known as the Bitter Springs formation were created and both were squeezed and buckled into tight curves during an intense period of mountain-building activity between 340 and 310 million years ago.

The late-afternoon sun bathes the Heavitree Range in a golden glow (opposite), making a stunning backdrop for a young whitewood rising from the rocky, spinifex-dotted plain. Like so much of the Red Centre, the West Macs landscape is at its most vivid around sunrise and sunset, so most daytrippers miss seeing it at its very best.

Sunlight paints the cliffs of the Heavitree Range and highlights the different plant communities that provide such a diversity of habitats for the rich wildlife of the West Macs. The mulga shrublands in the foreground gradually give way to spinifex grasslands then rocky slopes dotted with a variety of shrubs. They're among almost 600 plant species recorded in the park, which was gazetted in 1992.

Some of the finest scenery in the West Macs greets visitors who make the easy 15-minute walk to Ghost Gum Lookout, 70 m above the permanent waterhole at Ormiston Gorge. Afterwards they can cool off in the refreshing pool, which has an average depth of 12 m, or perhaps contemplate one of the many longer walks nearby.

Few of the park's 180,000 annual visitors glimpse many of these creatures because most rush through on day-trips along Namatjira Drive and don't venture more than a 10-minute walk from their car or coach. That's enough time to reach some attractions, including Simpsons Gap, where black-footed rock wallabies are easy to spot among the jumble of rocks beside the waterhole; Ellery Creek Big Hole and Glen Helen Gorge, superb swimming holes framed by imposing rocks and lush vegetation; and Ormiston Gorge, perhaps the most attractive of the waterholes. But you need time to explore some of the less accessible spots – like the ridge tops near Mt Giles, the nearby wet gorges and the quieter waterholes – and the more time you spend in the park, the more rewarding your visit will be.

Without a doubt the best way to enjoy the timeless grandeur of the West Macs is on foot – and the ultimate walk is the 220 km Larapinta Trail, mentioned earlier. But there are many other walks, ranging from easy half-day hikes to demanding two- or three-day treks like the one to the craggy summit of 1389 m Mt Giles.

If I had to single out one walk to recommend, it would be the panoramic Pound Walk from Ormiston Gorge. The park leaflet grades it moderate and suggests it can be completed in two or three hours; I found the walking easy but the scenery so spectacular and the wildlife so prolific that I took a full day.

As a change from walking, try cycling the Simpsons Gap Bicycle Path, which meanders for 17 km from the outskirts of Alice Springs through woodland to Simpsons Gap, or floating on an air mattress through spectacular Redbank Gorge; try swimming at Ellery Creek Big Hole or Glen Helen Gorge or exploring the 700-million-year-old red, yellow and white cliffs at the ochre pits, where generations of Aborigines mined the pigments used in their ceremonies and artworks. You can even explore some of the park by camel or from the air on regular scenic helicopter flights operated from Glen Helen Homestead, a historic and cosy lodge nestled beside the Finke River.

More than any other park in the Red Centre, the West Macs has something for everyone and despite its popularity, it's surprisingly easy to find solitude. On three of the four nights Tony and I camped in a clearing in the mulga scrub at Serpentine Chalet bush camping area, near the ruins of an early tourism venture, we were the only people there. And the 11 sites are so well spread out that even on the fourth night we couldn't see, or hear, our "neighbours".

Tnorala (Gosse Bluff)

A windswept trig point at Tyler Pass, near the western boundary of the West MacDonnells National Park, offers expansive views to the south of the ranges where the parched spinifex-covered hills drop to a vast flat plain that stretches to the horizon. Rising halo-like from the plain are the purple sandstone walls of one of the Red Centre's most remarkable natural features, Tnorala.

The weathered walls are the remains of a crater that was 20 km across and more than 2000 m deep when it was formed 130 million years ago by a comet that struck the

The breathtaking half-day Pound Walk at Ormiston Gorge takes walkers into the 10 km diameter pound, which is fringed by 200 m cliffs, and offers spectacular views of Mts Sonder and Giles before following shady Ormiston Creek back to the gorge. Euros, black-footed rock wallabies and a variety of birdlife add to the enjoyment.

German cyclists Susanne Kopf and Stefan Breyer (above) pause for a drink on the 17 km Simpsons Gap cycle path. Designed for both walkers and cyclists, the path winds across low hills and timbered flats between Simpsons Gap and John Flynn's grave, on the outskirts of Alice Springs. Pleasant camping spots like this one (right) at Redbank Gorge can be found throughout the West Macs and range from those with virtually no facilities, where campers can roll out their swags under the stars, to well-equipped sites at Ormiston Gorge and Glen Helen Homestead that offer campers and caravanners hot showers and flush toilets.

Guests at comfortable Glen Helen Homestead, which nestles in a bend in the Finke River, can admire this view (above) from almost every window. Date palms, possibly planted by Afghan cameleers, thrive among the native trees and grasses that line the river. Built on the site of a 1905 homestead that served Glen Helen station, the resort was one of the first in the Red Centre and has been welcoming tourists since the 1950s. In 1986 it was gutted by fire and shortly after being rebuilt, it was flooded when 400 mm of rain fell in less than 24 hours over Easter 1988. It is now under Aboriginal ownership. At the nearby waterhole that blocks Glen Helen Gorge (left), a towering sandstone cliff dwarfs swimmers preparing to take the plunge into what many visitors believe is the prettiest swimming spot in the West Macs. Aborigines believe it is guarded by a serpent.

From the distance, the weathered sandstone walls (top and above) of Tnorala (Gosse Bluff) could easily be mistaken for another mountain range, but they're the 4 km wide remnant of a huge crater formed 130 million years ago by the impact of a comet.

Earth with a force a million times more powerful than the Hiroshima bomb. Aborigines dispute this version of its origins, saying it was formed when a mother who was dancing across the sky put her baby aside in a wooden carrier that toppled over and crashed to Earth.

After viewing the crater from afar, it's well worth driving into the picnic area sheltered by mulga woodland inside the formation for a closer look at the brilliant orange walls that in places rise to about 100 m.

Few visitors to the West Macs venture to Tnorala because most retrace their route along Namatjira Drive to Alice Springs. A good alternative is to loop south then east along Larapinta Drive, which runs parallel to the southern flanks of the MacDonnell Ranges. You can continue along Larapinta Drive to Hermannsburg, Wallace Rockhole, Standley Chasm and back to Alice, or join the new Mereenie Loop Road, which provides access to Kings Canyon.

Hermannsburg

Driving into the historic precinct at Hermannsburg is like entering another world. A cluster of neat, whitewashed buildings – some fronted by manicured lawns, date palms and picket fences – surround a picture-book church that's partly shaded by two river red gums rising from the dusty red earth.

MELVA LOCKETT, MORTLOCK LIBRARY, STATE LIBRARY OF SA

Artist Albert Namatjira (left) was born in Hermannsburg in 1902 and drew his inspiration from the West Macs landscape around the settlement. After showing promise as an artist, he spent eight weeks at nearby Palm Valley with his mentor, Rex Battarbee. Shortly afterwards, in 1938, he sold every painting displayed in his first solo exhibition in Melbourne and went on to become one of our most famous artists, represented in major galleries and the private collection of Queen Elizabeth II. Two of his paintings are displayed in a gallery in the restored manse at the settlement, where Coralie Williams (below) discusses a Namatjira-style painting by Benjamin Landara with a visitor. Namatjira prints are available at the tea room, which also sells delicious Devonshire teas to raise money for restoring more of the 19th-century buildings.

In a sense, Hermannsburg is another world. The Lutheran mission was established in 1877 by German missionaries who spent nearly 20 months trekking overland from Bethany, in the Barossa Valley near Adelaide, in an incredible quest to "civilise" the Aborigines through the gospel.

Their bold vision suffered fluctuating fortunes and was abandoned for three years from 1891 until a new wave of missionaries ventured there. The mission became a thriving settlement with 500 residents, a cattle station, school and even a tannery where cattle and kangaroo skins were made into boots and rugs.

In 1982 the Lutherans handed the mission back to its traditional Arrernte owners and today the buildings, which date back to the 1880s and 1890s, are a major tourist attraction. Some are solidly built of locally quarried sandstone but others, like the old bakery, show the bush ingenuity of the early settlers. Half the building was fashioned from flattened kerosene tins and the remainder from corrugated iron.

Since 1987, many have been meticulously restored, including the little church, built in 1897, where there are regular talks about Hermannsburg's most famous son, Albert Namatjira, who was born in the settlement in 1902. Two of his distinctive watercolours are displayed in a small gallery in the manse that was built in 1888 for the first missionaries. Also on display are paintings by five of his children, four of his grandchildren and several other relatives.

Five-star camping (above). Shaded by river red gums and within a stone's throw of a perfect swimming spot, this idyllic bush camp on the Finke River near Glen Helen has long been a favourite of Alice Springs residents. Similar spots abound in both the East and West Macs, but – not surprisingly – locals jealously guard their exact whereabouts. At nearby Hermannsburg, potter Rachel Ungwanaka (right) displays two of the pots she has decorated with a design that depicts bush-tucker gathering. The distinctive work of the Hermannsburg Potters, as Rachel and her colleagues are known, is much sought after and is available in the settlement's tea room and at a number of galleries in Alice Springs.

Wallace Rockhole

Glenys Porter pointed to a large circle etched in the rock. "When you see a carving like that, it means there's water not far away and you keep looking," she said. "A small circle means it's further away – maybe 50 or 100 kilometres. There's always an animal track, like emu or goanna, and you follow that to the waterhole."

Glenys, shy and quietly spoken, was giving an Aboriginal survival lesson as we walked from the centre of Wallace Rockhole, a remarkable community 92 km west of Alice Springs, to the permanent waterhole from which it takes its name. Our path through a rocky gully in the James Range was dotted with mulga, bush tomatoes and witchetty bushes that would have provided food for the generations of Aborigines who visited the waterhole.

"While the men scooped up water, the women gathered grass and mulga seeds, which they ground with a large stone on a piece of rock," she explained, pointing to several hollowed rocks. The ground seeds were mixed with water and cooked in coals to make a nutritious bread.

Glenys has no idea how many Aborigines would have followed the track we were walking along, but a large gallery of red-rimmed hand stencils beneath a rock overhang a little further on indicated it was well used. "That's like a visitors' book," she said, pausing beneath the stencils. "It could be over a thousand years old."

In front of us was the rockhole – small, but a lifeline to Aborigines because it never dries up. Glenys assured me: "It's good water. You can drink it. The kids know not to swim in it because it's drinking water."

What's remarkable about Wallace Rockhole is that, unlike many Aboriginal communities, it positively welcomes visitors and Glenys and some of its 130 other residents enjoy sharing their culture during tours that often include sampling witchetty grubs, kangaroo tail and damper. "Tourism is important because it generates new dollars from outside the community," Glenys's husband Ken, the settlement's energetic administrator, told me later.

We were standing in the bright and airy arts and crafts centre where six people – including community president John Abbott – were working on paintings that would be sold in the adjoining gallery, which also sells wood carvings and ceramics. Ken said they were some of the 38 residents who earned their unemployment benefit through the Community Development Employment Program.

"We're looking at developing a market garden to grow dates, grapes and bush tucker for restaurants in town and we may develop a nature walk for visitors," he said. "Once we can get things up and running we can get more money."

Residents' pride in their settlement is obvious. The freshly painted homes and public buildings are surrounded by well-tended gardens and linked by streets neatly edged with white-painted stones. Oil drums have been recycled into brightly painted litter bins and there's not a scrap of litter or graffiti in what many people regard as a model Aboriginal community.

Wallace Rockhole resident Benjamin Abbott (above) prepares to serve cooked and skinned kangaroo tail to visitors during one of the regular bush-tucker tours that generate valuable tourist dollars for the community. Other attractions at the settlement are rock engravings and this gallery of hand stencils (below).

TONY STANTON

Tourists are dwarfed by nature (above) as they begin gathering in Standley Chasm an hour before midday to watch the sun pass overhead and paint its sheer walls a brilliant orange. Every lunchtime, the informal lounge at Glen Helen Homestead (above right) is a hive of activity as visitors enjoy a cool drink or sample the home cooking. The adjacent restaurant has won a string of awards and is considered one of the finest in the Territory.

Standley Chasm

Arrive at this dramatic slit in the MacDonnell Ranges near midday and you'll be greeted by the slightly disconcerting sight of several hundred people clogging the deep chasm, which is less than 5 m wide in places.

They're waiting for one of the Red Centre's many spectacular light shows – the effect of the sun passing overhead and painting the quartzite walls a fiery orange. It's a dazzling sight, but most visitors rush away afterwards and don't take in the beauty of the chasm, which is situated at the end of a luxuriant wet gully choked with ferns and cycads.

Travel wise

Namatjira Drive is sealed east of Glen Helen and Larapinta Drive is sealed east of the Wallace Rockhole turn-off. The remaining roads, including those to Gosse Bluff, Wallace Rockhole and the new Mereenie Loop Road to Kings Canyon, are unsealed but suitable for 2WDs.

Permits are necessary to use the Mereenie Loop Road. They are available free of charge

from the Central Land Council, Alice Springs Visitor Information Centre, the Ntaria Council Office at Hermannsburg, Glen Helen Homestead and Frontier Kings Canyon. A transit permit is needed for the trip to Tnorala and is available from the Central Land Council.

Fuel and food and/or meals are available at Glen Helen Homestead, Hermannsburg and Wallace Rockhole. Meals can be bought at Standley Chasm.

Glen Helen Homestead offers accommodation and camping, and camping is also available at Ormiston Gorge and Wallace Rockhole; there is basic camping (no water) at Ellery Creek Big Hole and Redbank Gorge, and bush camping at Serpentine Chalet. Facilities in the West MacDonnells National Park are being expanded, so for up-to-date information contact the Conservation Commission of the Northern Territory.

(Note: Food and meals may be of the most basic kind and travellers are advised to carry their own reserves. Accommodation varies from basic bunkhouses and on-site vans with shared facilities to motels with private facilities. Many places have a limited range. Camping refers to sites with showers, toilets and water. Most also have power available and are suitable for tents and caravans, access permitting. Basic camping indicates that a pit toilet and water are usually available; bush camping areas generally have no facilities. Camping fees apply at most sites.)

Motu Anderson (above) demonstrates the perfect way to cool off at Ellery Creek Big Hole. One of the most leisurely ways to explore the timeless grandeur of the West Macs is by camel (above left). Two-night safaris from Redbank Gorge to Glen Helen Homestead follow sandy river beds and rocky ridges that camels take in their stride. Trekkers spend their first night sleeping under the stars and on the second they swap their swags for comfortable beds.

In search of the past in the East Macs

The MacDonnell Ranges show a gentler face on the eastern outskirts of Alice Springs, but further east there are dramatic gaps and gorges that rival those of the better-known West Macs – yet attract only a fraction of the visitors. This view of Mt Undoolya, 25 km east of Alice Springs, shows the impact of plentiful rain on the

Crawling on my hands and knees, 6 m underground in the MacDonnell Range Reef Mine at Arltunga, I felt nothing but admiration for the miners who toiled here some 100 years earlier in their search for gold.

There was at least a metre of crawl space in the main tunnel, which had been hacked through the hard, white quartz with hammers and chisels, and sufficient for me to almost stand in places. But as I moved slowly along its 12 m length, trying to avoid stabbing myself on the sharp pieces of quartz littering the floor, I noticed that many of the side tunnels were less than 50 cm high.

I lay on my stomach and shone my torch down one, disturbing a tiny bat that flew within millimetres of my face. The tunnel didn't seem to get any bigger as it disappeared beyond my torch beam, yet miners had wriggled into these dark, cramped spaces every day for weeks on end to chip away at the gold-bearing quartz.

From 1887 Arltunga, 90 km east of Alice Springs, was the site of one of Australia's remotest gold rushes. Eager prospectors trekked huge distances to the rugged east MacDonnell Ranges, hoping to make their fortunes. The government established a battery, extraction plant and assay operations to process the ore, and a pub and store opened to cater for the residents of the Red Centre's first town.

But Arltunga's boom was short-lived. Its isolation meant living costs were high and the miners were hampered by extreme summer temperatures and a constant lack of water. Its population – and that of nearby Winnecke – peaked at 400 in 1903, but 10 years later the stores and processing facilities closed as the miners drifted away.

There's been sporadic activity since, but today Arltunga is a ghost town – and a monument to those early pioneers, who left behind a wealth of mine workings, buildings and machinery for visitors to explore.

Arltunga Historical Reserve

"It looks like hard work and believe me it is," senior ranger Kristen Appel said as she chipped rock from a quartz vein in the ceiling of the Great Western Mine with a hammer and chisel. "But there are some tricks to the trade. The gold occurs in cracks like these containing other minerals called pyrites."

Shafts of sunlight penetrated two huge rock windows in the semi-underground mine as I sat on

surrounding pastoral land. The same rain would have made the many creeks that cut though the ranges flow temporarily – some as torrential rivers – and highlights the boom or bust nature of the Red Centre climate, in which years of near drought can be followed by a year with more than twice the average annual rainfall.

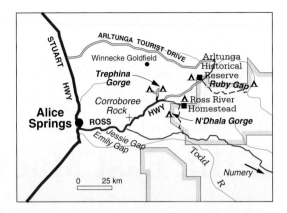

In the cool of the Great Western Mine at Arltunga, senior ranger Kristen Appel explains how miners obtained 17 kg of gold, worth $1 million today, from its quartz walls during an 11-year period at the turn of the century. Winter ranger-guided talks give a real feel for the life of the early miners and include demonstrations of gold panning and of crushing ore mechanically in a battery stamper and by hand in what's known as a dolly pot.

TONY STANTON

a ledge watching Kristen bring the early miners' lot to life during a regular winter ranger-guided tour to the mine, which was about the size of a lounge room. It was one of the most successful at Arltunga, having yielded about 17 kg of gold – worth about $1 million today – from more than 600 tonnes of ore that was chipped away and dragged out in buckets between 1899 and 1910.

Kristen explained that blasting was rare because it was expensive and dangerous. "One miner who used dynamite forgot the golden rule – that if it doesn't go off when it's supposed to, it could have a slow fuse," she said. "He waited and waited, and it finally exploded just as he went over to check it. A piece of rock the size of a bullet took off his left testicle."

The hapless miner was treated on the spot with mineral turpentine, then had to endure a two-week buggy ride to Oodnadatta and a train journey to Adelaide for further treatment. "He came back and reworked his mine, but after that the other men didn't use dynamite very much – and if they did they were very careful," she said.

Once Kristen had loosened some rock, she took it outside and placed it in a metal pot, called a dolly pot, to crush it. She explained that although the miners sent their ore to the government battery for crushing, they often dollyed small amounts to make sure gold was present before paying cartage and processing fees, which could swallow up nearly half what the gold was worth.

I watched as she tipped the crushed ore – resembling coarse sand – into a metal pan, added a little water and began gently swirling it. "As you twirl it, it develops a tail," she explained. "If you've got any gold, it'll stay behind in the tail because it's heavier." We spotted some fool's gold – pyrites – then the merest glint of the real stuff. Kristen quickly dashed my hopes. "That won't even be worth 2 cents," she said.

Although the two mines were the highlight of my visit because they brought to life the working conditions of the prospectors, there is a lot more to see at Arltunga and I'd recommend spending at least a full day exploring the 5491 ha reserve. It's best to begin at the excellent Visitor Centre, where displays and an audiovisual presentation give a fascinating insight into how harsh and isolated life was in the settlement's heyday. Afterwards, wander out to the nearby display of equipment, most of which has been retrieved from Arltunga and Winnecke.

The remainder of the site covers a large area amid steep, grassy hills that are strikingly different from tree-dotted slopes elsewhere in the East Macs. The miners chopped down all the mature trees for firewood and shelters – and in doing so created a harsh landscape for themselves and those who followed. The only trees sprouting from the parched tussock grass are the occasional young ghost gum and corkwood, and small mulga, cassia and wattle bushes that offer little shade.

The most extensive ruins are those of the government works, dating from 1898, where the staff running the processing operations lived and worked in comparative luxury. The stone, two-roomed manager's residence, with its shady verandas front and back, contrasts starkly with the remains of miners' huts dotted about elsewhere.

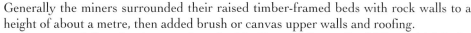

TONY STANTON

Generally the miners surrounded their raised timber-framed beds with rock walls to a height of about a metre, then added brush or canvas upper walls and roofing.

Be sure to follow the short track to the MacDonnell Range Reef Mine, which is the only underground mine at Arltunga. Ladders at each end give access and you'll need a torch if you plan to crawl underground – and you should, to get a feel for what an early miner's life was like. The mounds of dirt beside the track are the remains of alluvial workings. Before rich seams of reef gold were discovered in the area in the late 1880s, miners spent several years digging in creeks and gullies for alluvial gold.

The Great Western Mine, which is cut into a hillside, is reached by a track beyond the Crossroads area – once the hub of the township with a pub, store and the first of two cemeteries. The second cemetery, White Range, is the more interesting. Its 12 bush graves include that of Henry Luce, Arltunga's most successful miner, who made an estimated $2 million in today's money from the goldfields.

The cemetery looks across to the White Range Mine, where open-cut mining began in 1990 but ceased soon afterwards – dashing the hopes of many modern-day prospectors and proving that mining at Arltunga is no easier today, despite the advances in technology and communications.

Ruby Gap Nature Park

Alice Springs locals cherish this rugged and remote 9257 ha park where the Hale River has carved a dramatic, winding gorge through the ranges, and as soon as Tony and I

Arltunga is undoubtedly the Red Centre's most fascinating historic site and one of the finest examples of an early goldmining town anywhere in Australia. The eager prospectors who trekked vast distances from the depressed southern goldfields from 1887 found a harsh environment with little water, extreme temperatures and high living costs – and within 20 years many had moved on to seek their fortunes elsewhere. Among the many relics they left is a wood-fired boiler (above), which produced steam to power the engine that drove the ore crusher in the government works. The nearby police station and jail, which have been extensively restored, were built in 1912 after police complaints about working conditions at the settlement. A policeman had been stationed there from 1899 and until the station was built, he made do with a single room that served as an office, home and lock-up. Any prisoner had to be chained to a leg of the policeman's bed!

TONY STANTON

reached it – after a bone-jarring 46 km drive over the rocky, corrugated track from Arltunga – I could see why.

The wide, sandy bed of the Hale was bordered by velvety grass banks and lined on each side by sprawling river red gums that reached up to 50 m. Dwarfing everything were spectacular, craggy, red rock walls that were thrust up, faulted and folded more than 300 million years ago.

We followed the track upriver for about 2 km until we reached a perfect camping spot in a small clearing in mulga and bloodwoods above a bend in the river. Towards sunset we clambered up a gentle hill behind our camp and watched the sun's last rays igniting the rugged landscape that surrounded us.

The following morning at dawn we walked along the soft, sandy river bed for more than an hour to Glen Annie Gorge, where bulrushes fringed a series of shallow waterholes and stunted ghost gums poked out of crevices in the high red cliffs on either side.

There's usually a deep, permanent waterhole blocking the gorge, but rain a few months earlier had washed in a lot of sand – and hadn't been heavy enough to scour it out again. The lack of water made it easy to continue upstream to find one of the Red Centre's earliest European graves, that of a fortune seeker called Fox who shot himself after an 1886 ruby rush went sour.

Explorer and surveyor David Lindsay had sparked the rush when he found what he thought were rubies as he dug for water in the sandy bed of the Hale. Eighteen months later, word came back from London that the so-called rubies were in fact garnets – beautiful but almost worthless. It was then that alluvial gold was found in Paddy's Rockhole in Arltunga, so many of the prospectors headed there. Lindsay, meanwhile, was commissioned to survey the township of Stuart, now Alice Springs, to supply the growing settlement of Arltunga.

On our way through the gorge we passed dingo and rock-wallaby tracks – and we'd seen zebra finches, babblers and honeyeaters earlier. But there were no other people – we'd been alone that night in Ruby Gap. We were so captivated by our surroundings that we stayed a second night, relishing the wilderness atmosphere of this imposing gorge. Ruby Gap had proved a real gem.

The Ross Highway

The route to Arltunga, Ruby Gap and other points east of Alice Springs is along the Ross Highway, which is named after Overland Telegraph Line surveyor John Ross, and runs parallel to the rocky southern flanks of the east MacDonnell Ranges.

Scenically, the East Macs tend to be eclipsed by the more spectacular and higher West Macs, yet they abound in natural attractions and Aboriginal and European history.

The big advantage of the East Macs is that few tourists venture there – Arltunga attracts just 14,000 a year and even Trephina Gorge, with scenery to rival the best of the West Macs, attracts only 55,000. Perhaps that's why the wildlife is often easier to see.

Imposing Corroboree Rock (above), a dolomite outcrop, is one of many sites in the east MacDonnells that is significant to the Eastern Arrernte people, who used it to store sacred objects. Equally dramatic is Glen Annie Gorge (opposite) in Ruby Gap Nature Park, where the Hale River has carved a wide chasm through towering quartzite cliffs as it makes its way to the Simpson Desert. Explorer David Lindsay, who named the gorge after his wife, sparked a brief ruby rush there in 1886 when he discovered glittering red stones – later identified as almost worthless garnets – in the sandy bed of the Hale. The park now offers some of the finest bush camping in the region.

The bustling bar (above) of Ross River Homestead dispenses old-fashioned outback hospitality along with cool drinks and hearty tucker, and is an ideal place for visitors to meet some of the locals. At nearby N'Dhala Gorge Nature Park (below), delicate circle and feather engravings believed to be thousands of years old tell a Dreamtime story of two caterpillars. They are among nearly 6000 engravings that decorate the rocky gorge.

I've only seen emus once in the Red Centre, and that was beside a creek in the East Macs, and I've had my best sightings of black-footed rock wallabies there.

Heading east from Alice, you'll soon reach Emily and Jessie gaps, the first of many water-worn gaps in the ranges. Both are pretty rather than spectacular, but they're important to Aborigines because they're part of a Caterpillar Dreaming trail that runs south-east to the edge of the Simpson Desert.

A little further east, a walking track runs through native figs, bloodwoods and whitewoods that surround Corroboree Rock, a distinctive dolomite outcrop where Eastern Arrernte elders explained their sacred objects to the younger men. They stored the objects in a cave near an easily visible window in the rock but, despite the name, they probably never danced there.

Ross River Homestead, 64 km east of Alice Springs, is one of the Red Centre's oldest surviving tourist ventures, dating back to 1958. Earlier this century its bustling, beamed bar, popular with both locals and tourists, was the hub of a thriving horse-breeding venture. The station owners switched to cattle when demand for their horses slumped after World War I, and many of the horses were allowed to run wild. Their brumby descendants still roam the area.

Trephina Gorge Nature Park

The main feature of this rugged but easily accessible 1770 ha park 53 km east of Alice Springs is a large, semi-permanent waterhole that sits in a bend in the wide, sandy bed of Trephina Creek and is dominated by towering red quartzite rock. There's plenty of shade provided by river red gums and ghost gums, and it's a great place to relax and watch the prolific birdlife that includes water-frequenting birds like black-fronted plovers, Pacific herons and fairy martins, as well as noisy residents of the open woodland like galahs and red-tailed black cockatoos.

Early mornings and late afternoons are good for spotting the many black-footed rock wallabies that live in the area, although you'll often see them relaxing in shady caves high above the gorge in the middle of the day. I've seen more of these agile rock dwellers here than anywhere else in the Red Centre.

There are several walking trails and you don't have to venture far along them to enjoy a real sense of isolation. Be sure to follow either the Gorge Walk or the Panorama Walk, which offer great views of the gorge and waterhole. Both are easy walks that can be completed in less than an hour. On the way, look out for the holly-leaf grevillea with its distinctive red, lantern-like flowers.

If you're feeling more energetic, the Ridgetop Walk runs along the rugged main range to John Hayes Rockhole, a tranquil waterhole framed by red rocks. The walk takes a good four hours one way, so it's a good idea to arrange to be picked up or dropped off at the rockhole, which is also accessible via a rough 4WD track. The shorter, but equally rugged, Chain of Ponds Walk from the rockhole offers more great views.

N'Dhala Gorge Nature Park

This rocky gorge choked with ghost gums and native figs contains nearly 6000 Aboriginal petroglyphs or rock carvings – but only the most observant visitor is likely to spot many of the geometric symbols and animal tracks that have been etched into large, flat rock surfaces.

Even if, like me, you miss many of the engravings, the walk along the gorge is still enjoyable because it attracts few visitors. On the way, look for native tomatoes and for the varied birdlife that includes melodious grey shrike thrushes, colourful budgerigars and shy dusky grasswrens.

Travel wise

The Ross Highway is sealed for 70 km east of Alice Springs. The sections to Trephina Gorge and Arltunga are unsealed but suitable for 2WDs. The track from Arltunga to Ruby Gap is 4WD only and unsuitable for towing. A 4WD is recommended on the tracks to N'Dhala Gorge and John Hayes Rockhole.

Fuel, meals, accommodation and camping are available at Ross River Homestead.

There is camping at Arltunga Outback Tourist Park, basic camping at Trephina Gorge, John Hayes Rockhole (no water) and N'Dhala Gorge (no water); and bush camping at Ruby Gap.

Rising from a carpet of old-man saltbush (above), this dazzling ghost gum at Trephina Gorge Nature Park is believed to be the largest specimen in the East Macs at 33 m high and with a girth of 4 m. It's probably at least 300 years old and would dwarf its spindly neighbours at nearby John Hayes Rockhole (above left). Named after the region's oldest pastoral family, the tranquil rockhole is a favourite watering point for the area's prolific birdlife and can be reached by 4WD or by following the rugged Ridgetop Walk from Trephina Gorge.

In the footsteps of Giles

Sudden changes in the Red Centre weather can have a dramatic impact on its landscape and show its many rock formations in a very different light. This dust storm that almost obliterated the view of the awesome domes of Kata Tjuta from the nearby lookout blew up in the middle of a day that dawned still and clear – and quickly reduced visibility to a few metres. Such storms are not uncommon,

The glassy blue surface of long, reed and bulrush-fringed Boggy Hole was peppered with ducks and waterbirds as I crept down its soft, grassy western bank. I sat near the water's edge, camouflaging myself behind a pocket of reeds, and picked up my binoculars, knowing that, as a keen bird-watcher with a fascination for waterbirds, I had a treat in store.

Within 15 minutes I'd spotted black ducks, pink-eared ducks, grey teal, little pied and little black cormorants, Pacific herons, coots, dusky moorhens, purple swamphens, wood ducks, grebes and black-fronted plovers – as wide a range of waterbirds as I'd expect to see virtually anywhere in Australia, and far more than I hoped to see in its arid heart. Along the edges, where slender young river red gums dotted the wide valley between orange-red sandstone bluffs, were babblers, fairy martins, black cockatoos, honeyeaters and budgies.

More than 200 species of birds have been recorded at Boggy Hole, a magnificent permanent waterhole on the Finke River 113 km south-west of Alice Springs. I sat for more than an hour watching some of them as they fed and preened in the late afternoon sun, feeling pleased that Tony and I had chosen the 4WD Boggy Hole track as

our gateway to Ernest Giles country and Watarrka and Uluru–Kata Tjuta national parks, instead of the longer but quicker Stuart Highway route.

Giles camped at Boggy Hole on 30 August 1872 on his way north up the Finke and noted: "Several hundreds of pelicans in a large flock made their appearance upon the waterhole near the camp this morning; but no sooner they discovered us than they made off, before a shot could be fired at them."

I didn't see any pelicans but I did find the ruins of a police camp established in 1889 in a vain attempt to ease the growing tension between pastoralists and the Aborigines they were displacing. It closed a few years later after the officer in charge, William Willshire, was charged with the murder of two Aborigines. He was later acquitted and today little remains of this lonely bastion of law and order.

The following morning I was up at dawn for more birdwatching and was just about to pack up to head south when I noticed that the birds on the far side of the waterhole were becoming agitated. I grabbed my binoculars and saw a dingo and a huge feral cat confronting each other on the grass near the water's edge.

particularly after long periods of drought, and the effects of this one were intensified by a fire that had left the landscape littered with the blackened remains of spinifex and desert oaks, which were whipped up by the wind. Although some visitors were disappointed not to get a clearer view of one of the region's best-known landmarks, those with time to linger relished the mood the storm created.

Luxuriant Boggy Hole (above), one of several permanent waterholes on the Finke River, is a birdwatcher's paradise where more than 200 species have been recorded. Ernest Giles camped there in 1872 and although most of the lofty river red gums that would have shaded him were uprooted in 1988 floods, saplings are growing fast and it's still a delightful camping spot. Debris from the flood, when the Finke peaked at 7 m, can still be seen high in the branches of many of the surviving trees on the 4WD track between Larapinta Drive and Boggy Hole. The graceful red cabbage palms at nearby Palm Valley (opposite) seem even more out of place in the arid Red Centre than Boggy Hole. Also called Palm Valley palms, they're remnants of a rainforest that thrived when the region was much wetter, and survive because their shallow roots tap into a small, permanent water supply.

The cat's back was arched and it looked nearly as big as the dingo as they sized each other up. As the dingo tentatively edged closer, the cat hissed and lashed out so violently with its claws that I wouldn't have been surprised if the dingo had retreated. But it stood its ground, gradually edging the cat into the water, where it struggled as it began to sink into the soft mud. Finally, when the cat was obviously having difficulty keeping its footing, the dingo sank its teeth into the cat's neck and, a few seconds later, carried the limp body into the bushes away from the attentions of some circling crows.

I knew that, thanks to the wily dingo, life would now be a little safer for the birds that had given me so much pleasure.

Finke Gorge National Park

Boggy Hole is probably the least-known attraction of this 46,000 ha park carved over the millennia by the meandering of the 250-million-year-old Finke River, believed to be the oldest in the world, as it begins its 690 km journey to the western fringe of the Simpson Desert.

The best-known feature is Palm Valley, 18 km to the north-west, where first-time visitors can be excused for thinking they've been transported 1000 km north to the tropics. The spectacle of 12,500 stately red cabbage palms, some up to 25 m tall, creating a lush oasis in a long, narrow sandstone valley is remarkable.

The slender-trunked palms, which are up to 300 years old and appear to sprout from the reddish-brown sandstone, are remnants of a rainforest that once covered the area. The key to their survival is the small but permanent water supply. When rain falls, it soaks into the porous sandstone hills and filters down into the valley where the trees' shallow roots tap into it.

The best way to see the palms, which are found nowhere else on Earth, is to take the 5 km Mpulungkinya Track that winds through the trees then up onto the sandstone plateau above them for views down into the sheltered valley.

Although Palm Valley and Boggy Hole are fairly close to each other, the ruggedness of the terrain means there's no direct link between the two and drivers have to backtrack to Hermannsburg. For this reason, most of the park's 65,000 annual visitors see just the Palm Valley section, also taking in the cycad-filled gorge and the magnificent sandstone amphitheatre they pass on the way.

The majority rush through in a day, but I'd strongly recommend an overnight stay in the grassy, tree-studded camping ground beside the narrow bed of Palm Creek. It's as attractive a camping area as you'll find in any of the Red Centre's national parks and is regularly visited by colourful Port Lincoln ringnecks and melodic pied butcherbirds – and it has the added luxury of solar-heated showers.

I left the camping area at first light to walk to Kalarranga Lookout, where I sat watching the sunrise wash over the weathered sandstone walls of the vast amphitheatre with a golden glow. From there I headed along the spectacular 5 km Mpaara Track, which follows the Finke River then winds up and over the amphitheatre walls.

The Boggy Hole track

Driving down the sandy, occasionally stony, beds of Ellery Creek and the Finke River as they cut a swathe through the high walls of an ancient weathered sandstone gorge towards Boggy Hole, it's hard to picture the landscape when the rivers are flowing, as they do – briefly – in most years.

Yet all around is evidence of the last big flood over Easter 1988, when 40 cm of torrential rain fell in 24 hours and the Finke peaked at 7 m. Dozens of river red gums were uprooted as the river became a raging torrent and they left a trail of destruction as they were carried downstream, clearing everything in their path.

Signs are still visible high in many of the surviving trees – sometimes 5 m from the ground. Bundles of twigs and other debris may look like birds' nests at first glance, but most became snagged in the trees during the flood and will remain there until they're dislodged by fierce winds or more floods.

Be sure to stop about 6 km south of the park entry gate at a large, shady waterhole, fringed by lush grassy banks, where Ellery Creek meets the Finke. It's the start of the Glen of Palms, named by Giles. He recorded: "I was literally surrounded by fair flowers of many a changing hue … They alone would have induced me to name this Glen of Flora, but, having found in it also so many stately palm-trees, I have called it the Glen of Palms."

Giles found his journey up the Finke on foot and horseback hard work, but today's 4WDs make it fairly plain sailing. You'll encounter a number of soft, sandy patches where you need to keep momentum up, but drivers with some experience of sand should have few problems. However, these same stretches become like quicksand after the river flows, and vehicles can sink up to their wheel arches.

About 17 km south of Boggy Hole, the track leaves Finke Gorge National Park and passes Running Waters station. Later it parts company with the Finke and runs across vivid orange-red sand dotted with desert oaks. A little further on is a turn-off to Illamurta Springs, the site of another remote police camp that operated from 1893–1912 and later became a ration centre for dispossessed Aborigines.

The track then follows a 28 km seismic line that goes straight up and over a magnificent stretch of 10 m high sand ridges. The driving is generally reasonable – the one dune that did give us trouble was easy to bypass – until you reach the wide, sandy bed of the Palmer River, which requires care. Once at the Ernest Giles Road, it's 136 km to Watarrka National Park. The first 36 km is badly corrugated, but the rest is sealed.

Watarrka National Park

I was feeling exhilarated after completing the spectacular 6 km Kings Canyon Walk around the rim of the canyon when I met a rather dour Scotsman. "Is there much to see?" he asked. I told him I reckoned it was one of the finest walks in the Red Centre and he replied: "I've just wasted an hour walking along the creek. There was nothing to see."

Computerised records of the wildlife and plants of Finke Gorge National Park – including its butterflies and dragonflies (above) – are kept by ranger-in-charge Dennis Matthews on one of the most sophisticated Geographic Information Systems in Australia. Three years ago, more than 2000 brumbies were removed from the park to protect the fragile vegetation and the native animals that rely on it for food and shelter – and Dennis hopes the park's diversity will improve as a result. The appearance of Boggy Hole, a favourite watering spot for the brumbies, has already improved significantly. The dramatic gorge carved by the Finke River near its junction with Ellery Creek (opposite) is as impressive from the air as it is from the ground. The ancient river, which flows briefly most years, snakes 690 km across the Red Centre from the MacDonnell Ranges to the Simpson Desert and is believed to have followed the same course for at least 100 million years.

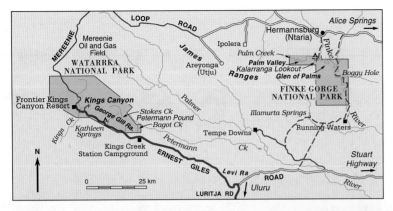

The steep climb to the top of Kings Canyon rewards walkers with spectacular views of Watarrka National Park's rugged sandstone landscape and the distant desert sand plains. Members of the Horn Scientific Expedition who visited the area in 1894 judged it to have the region's best scenery, animal and plant life, and much of it can be observed on the half-day walk around the canyon rim.

I'd done the same walk the previous afternoon, clambering along the rocky creek bed through shady river red gums and white cypress pines to the interior of the massive canyon. The late afternoon sun had added a brilliant glow to the sheer red walls, which rise to 150 m, and I'd watched hundreds of red-tailed black cockatoos flying overhead. True, there were no spectacular views – you have to work a bit harder for them – but I certainly didn't feel I'd wasted my time.

Giles didn't either when he saw the canyon in October 1872. "It appeared very red and rocky, being composed of enormous masses of red sandstone, the upper portions of it being quite bare, with the exception of a few cypress-pines moored in the rifted rock." He was so captivated that he spent a day exploring its waterholes and admiring the cycads that grew among the rocks. Three years later he applied for – and was granted – 2525 sq. km of pastoral land in the George Gill and Levi ranges – including the whole of Watarrka. The land, which he disposed of in 1883, is now part of Tempe Downs station.

I felt sorry for the Scotsman. I'll never know if he did the canyon walk, but I hope he did. It's a must for every reasonably fit visitor to Watarrka.

My own walk had started in the cool of early morning with a steep climb to the top of the canyon, where I paused to catch my breath and gaze across the expansive desert oak and spinifex-dotted sand plains to the distant George Gill Range, which Giles named after his brother-in-law. It's one of the finest views in the Red Centre and even the recent construction of a $16 million resort complex in its midst has had little visual impact.

My next stop was near the amazing orange beehive formations of the aptly named Lost City. Some 350 million years ago, these domes were cube-shaped blocks laced with horizontal hairline cracks. Since then, erosion by water and wind-blown sand has weathered them into their distinctive beehive shapes and created a sheltered refuge for some of the park's 60 rare or relict plant species and 100 reptile species.

Later I passed luxuriant ferns and cycads as I made my way into the oasis-like Garden of Eden, where Kings Creek runs through a deep gorge that ends in a large, permanent waterhole. Many walkers have to skip the short detour to the "garden" to meet their hectic tour schedules, but for me it's one of the highlights of the walk and the perfect morning tea spot. During the 40 minutes I relaxed there, I didn't see another person in its moist, shady confines.

Back in the dazzling sunshine on the main track, it was a different story. Dozens of people were walking around the rim, stopping regularly to gaze at the sheer red and cream walls, as I made my way around and down it, ending on the opposite side to where I'd started. My leisurely walk had taken little more than four hours.

Throughout the walk I noticed a lot more people than when I first did it three years earlier. Later, chief district ranger Ian King told me Watarrka is one of the highest-growth parks in the region, with visitor numbers rising steadily by 15 per cent a year over the previous three years to 100,000.

Walkers cross one of the new bridges installed to make the Kings Canyon Walk safer for the growing number of visitors following the 6 km track. Bridges and steps also help to protect the many rare and fragile plants that are among the 600 species found in the park.

Early-morning walkers are dwarfed by the sheer 150 m cliffs of the southern face of Kings Canyon (left). Huge slabs of red, iron oxide-stained Mereenie sandstone have fallen into the canyon as the lower layers of softer rock have gradually eroded away. Nestled among shady desert oaks, ironwood and mulga, Kings Creek station (above) was the setting for AUSTRALIAN GEOGRAPHIC's 1992 scientific expedition and is an ideal base for exploring Watarrka National Park and the neighbouring ranges. Owner Ian Conway has made something of a name for himself as Australia's most successful camel catcher and he and his wife Lyn are always happy to share their knowledge of the surrounding area. Visible from 50 km away, Mt Connor (opposite) is the only major landmark between Kings Creek and Yulara, and many visitors mistake it for Uluru. It's on Curtin Springs station land but can be visited on organised tours from Yulara.

To cater for that growth, more walking tracks are being developed. These range from the short, easy Kathleen Springs Walk, which is suitable for wheelchairs and leads visitors past former stockyards and cattle-watering points to a tranquil waterhole where the rare creeping swamp fern grows, to longer walks. Among these is a two-day walk from Kathleen Springs to the canyon that passes several Aboriginal art sites and will offer an insight into Aboriginal culture. A second canyon within the park may also be opened up to visitors to spread the load, and Ian says rangers are always happy to suggest routes for overnight walks to experienced walkers.

A major reason for the rapid growth in visitor numbers at Watarrka was the opening in 1991 of Frontier Kings Canyon, which caters for campers, backpackers and those seeking luxury motel accommodation. The resort is also the base for Kurkara Tours, an Aboriginal-run tour company that offers a number of innovative cultural tours.

Unfortunately the tours operate subject to minimum numbers and didn't run during my stay. But Kit Whillock, the trainer/coordinator, told me: "The whole essence of the tours is that they are very relaxed and personal. The Aborigines are very enthusiastic about them – particularly the longer tours because they're naturally shy and like to have time to get to know the people they are dealing with."

Among the tours are tailor-made walking expeditions with helicopter support. "We can structure a walk to fit in as many highlights as we can in the time people have available," said Kit. "Walkers carry a day pack, and a helicopter drops swags, food and a cook for the evening meal."

For those looking for a quieter alternative to the resort, Kings Creek station, 32 km further east, is worth checking out. It's run by Ian and Lyn Conway, whose intimate knowledge of the area contributed greatly to the success of AUSTRALIAN GEOGRAPHIC's fourth scientific expedition, which the Conways hosted in 1992. Ian introduced the 28 expeditioners to rugged sandstone country in the nearby George Gill Range near Bagot and Stokes creeks – and many said it was the most beautiful they'd seen in the region.

The spacious camping ground, developed in bushland dotted with desert oaks, ironwood and mulga, is probably the most attractive and natural in the Red Centre and is a firm favourite with AG Editor Howard Whelan. "I reckon it best captures camping in the outback," he said. "It caters for a lot of people but it manages to make them all feel they are in a remote bush camp."

The drive to Yulara

The drive from Watarrka to Yulara via the Ernest Giles Highway, Luritja Road and Lasseter Highway passes some of my favourite Red Centre scenery. It's desert, but there's not a bare dune in sight. The vibrant orange-red sand is covered with spinifex and a mass of other vegetation – from the delicate purple flowers of the parakeelya to the needles and Kiwi fruit-sized cones of the stately desert oaks that are at their most prolific in this area.

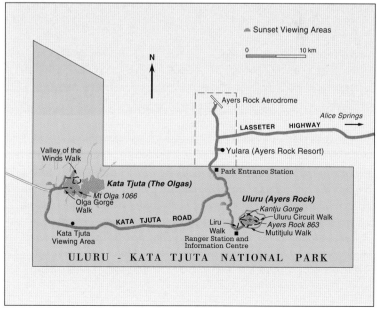

Ultra-modern Yulara (above) was created to accommodate tourists to Uluru and the people who look after them – and to minimise their impact on the rock – and it does so in award-winning style. Although it offers the only official accommodation within sight of Uluru, some tour companies and visitors prefer bush camps like this one – about 11 km east of Yulara and outside the national park – where English visitor Andrew Corner, on the left, watches sunrise light up the rock while the rest of his group catch up on sleep. Watching the sunset ignite the rock at the end of the day (following pages) is always a much less intimate affair.

Not long after joining the Lasseter Highway to Uluru, the blue-grey outline of a huge flat-topped rock looms on the south-western horizon. It's easily mistaken for Uluru at first glance, but it's Mt Connor, the least known of the Red Centre's three great tors (the others being Uluru and Kata Tjuta).

The sandstone mesa, which rises more than 300 m from the plain, is on private property and the only way to get close to it is to join an organised tour. But there are good views of it from a rest area a little further west, where steps lead to the top of a dune that overlooks a vast, shimmering salt lake on the other side.

Yulara

After kilometre upon kilometre of seemingly endless desert, the sight of Ayers Rock Resort and the ultra-modern tourist town of Yulara rising from the sand comes as a shock. Designed by award-winning Australian architect Philip Cox, it's architecture on a grand scale – akin to the Sydney Opera House – and closer inspection reveals it's also incredibly practical.

No part of the village is higher than the surrounding dunes and the eye-catching fabric sails and extensive bullnosed verandas provide much-needed protection from the harsh desert sun. The buildings are painted in muted pinks, reds, yellows and ochres to blend with their surroundings, and extensive use is made of solar panels, which provide about 70 per cent of the resort's hot water.

If you want to spend more than a day in the area (and you certainly should), you'll have to stay there – be it in the grassy camping area, the backpacker and self-catering accommodation or one of the luxury hotels. Yulara was created to cater for the growing number of tourists visiting the rock, which is 12 km further south, and to minimise their visual impact on it – and it's done so impressively since it opened in 1984. It's a well-oiled machine accommodating some 320,000 people, more than two-thirds of them from overseas, every year and offering a wide range of experiences to suit even the most esoteric tastes.

You can take a scenic flight over Uluru and the surrounding area in a helicopter or small plane, travel around the rock in a chauffeur-driven limousine or on the back of a Harley Davidson motorbike, or join one of a multitude of coach or walking tours. You can learn about the dazzling night sky, join occasional painting or photography courses (some of the latter are led by Mike Langford, AUSTRALIAN GEOGRAPHIC's 1992 Photographer of the Year) or even spend time living with Pitjantjatjara people and taking part in traditional food-gathering and ceremonies at one of two settlements just over the South Australian border. The choice is endless.

Even if you choose not to stay at Yulara, set aside an hour or so to tour the Visitor Centre, which has excellent displays on the desert landscape and wildlife, and to wander through the town centre. As well as catering for tourists, it's home to 800 resort staff, tour operators and support staff, and has shops, fire and medical services, a school, community centre and government offices.

Uluru (Ayers Rock)

Maureen Natjuna talked loudly and animatedly in her native Pitjantjatjara tongue, her round face regularly breaking into a toothy grin. Her brightly printed skirt and blouse and the Eagles beanie pulled over her black hair were symbols of the 20th century, but she was giving a demonstration of a 30,000-year culture.

Maureen was leading me and a group of 20 tourists from four continents on the Liru Walk, a gentle 2 km stroll from the ranger station to the base of Uluru. She was one of a number of senior women from nearby Mutitjulu community who devised the walk to give visitors an insight into the culture of Anangu, as the traditional owners are known. Since Anangu assumed a key role in the management of World Heritage-listed Uluru–Kata Tjuta National Park in 1985, they've discouraged visitors from climbing Uluru, which is sacred to them. Instead they urge them to see it from their perspective and to learn about the culture of the original inhabitants through a series of guided walks.

Two Anangu trainee rangers and a non-Aboriginal senior ranger were on hand to translate for Maureen as she led us to a brushwood *wiltja* (shelter), where she gave a demonstration of spinifex threshing.

I watched as she hit handfuls of spinifex with a mulga branch to loosen the yellow crystals of resin at the base of each stem, sifted them in a *piti* (bowl) and heated them with a flaming twig. As they began to melt, she worked them into a ball that she put on the end of another twig and held over a fire. Then, using her fingers and the occasional bit of saliva to stop her skin from burning, she worked it into *kiti*, a pliable mass that

would have been used for bonding stone tools to wooden handles. She also demonstrated how women used fire to sharpen and harden the digging sticks they used to unearth witchetty grubs and honey ants. The stick was simply held over the flames and the charred bits scraped off.

Further on, near a huge bloodwood, Maureen showed us how an oval piece of bark would have been peeled from the tree to make a *piti*. As well as being used for gathering and preparing food, *piti* were used to carry water and even babies. A ring made from emu feathers and human hair was worn on the head to carry the *piti* when collecting water.

Beside a mulga tree, Maureen demonstrated how to make a *miru* (spear thrower) from mulga bark and pointed out the long, slender branches of the spearwood tree from which spears were fashioned. Then, much to her amusement, she gave a demonstration of spear throwing – considered to be a male skill.

Firelighting was also a man's job, so at our next stop Maureen got two male volunteers to hold each end of a *miru* and rub it to and fro across a groove in a mulga stick. The groove was packed with rabbit droppings and the friction should have caused

Designed by the Anangu, Uluru's traditional owners, the popular Liru Walk (above) offers a fascinating insight into an ancient culture. Here ranger James Eldridge translates from the Pitjantjatjara used by Maureen Natjuna and Kakali Kinyin as they demonstrate the art of making KITI, a bonding material, from spinifex resin. Other tourist options are watching sunrise gently wash over the rock and the desert oak-studded sand plains (opposite left) and the nightly Sounds of Silence dinner (opposite right). Guests enjoy a gourmet feast that includes emu and camel as the sun sets, then relax over coffee and port under a star-studded sky. An astronomy expert gives a talk on the night sky, then offers diners a closer look through his powerful telescope.

For many visitors to the Red Centre, climbing 348 m Uluru is the highlight of their stay. An increasing number, however, are heeding the request of the traditional owners, the Anangu, and are not climbing the rock because it is sacred. Out of respect, they are following some of the many guided walks around its base instead. The 9.4 km Base Walk takes 3–4 hours and includes the opportunity to see some ancient art sites.

them to smoulder enough to ignite the dry grass or other kindling that would be applied. Our volunteers were unsuccessful – but Maureen told them Aborigines usually carried a firestick, of slow-burning mulga, to save them having to start from scratch.

As we neared the base of the rock, Maureen discussed *Tjukurpa* (Anangu knowledge), which explains the spiritual relationship between the land, the traditional law and the people. The walk had taken 90 minutes, the average time it takes someone to climb the rock, it hadn't cost a cent (although we did have to book in advance) and it had given me a new insight into the desert Aborigines' resourcefulness and the depth of their knowledge. A straw poll among my fellow walkers revealed that everyone was as impressed as I was by what we'd seen and heard.

Two days earlier, I'd joined senior Anangu ranger Leroy Lester on the daily Mala Walk, which starts at the base of the climb and runs clockwise for 2 km past caves and art sites to Kantju waterhole, one of the few reliable sources of water around the rock. The following day I'd walked the 9.4 km circumference of the rock with guide Ian Archibold of Uluru Experience. On both walks I'd been encouraged to view the rock from an Anangu perspective and had its many features interpreted.

To Anangu, every nook and cranny of Uluru's weathered surface has a meaning that is interpreted through stories. One of the most graphic is of the battle between Kuniya, the sand python, and Liru, the venomous snake. Ian pointed out the marks on the rock where Kuniya moved in to strike Liru with a *wana* (digging stick) and where the first blows fell. Later we saw the piece of rock that represents Liru's shield falling to the ground as he was fatally wounded, and the nearby spearwood bushes poisoned by Kuniya's rage.

Sadly, many visitors shun these activities and instead concentrate on climbing 348 m Uluru. When Ernest Giles visited the rock in 1873, he commented: "The appearance and outline of this rock is most imposing for it is simply a mammoth-monolith, that rises out of the sandy soil around, and stands with a perpendicular and totally inaccessible face at all points, except one slope near the north-west end, and that is at best a most precarious climbing ground."

The installation of a handrail in the 1960s made it a little less precarious, but the climb is still hard work and 26 people have died since 1965. Each year a similar number have to be rescued. If you must climb it (and the Anangu ask that you don't, because the rock is sacred), wear sturdy shoes and loose protective clothing, carry – and drink – plenty of water, and start early to avoid the heat of the day. Don't consider climbing if you suffer from a heart condition or breathing difficulties.

In deference to Anangu wishes and also because I don't have a head for heights, I haven't done the climb. But I have taken part in what is probably Uluru's major tourist activity – sunset viewing of the startling colour changes.

The spectacle of the rock changing from a pinky-blue to fiery red and then a burnished orange is among the most impressive in the Red Centre. Of course, the rock's not actually changing colour. What causes the apparent change is the reflection of the

No two Uluru sunsets are the same and every evening hundreds of visitors assemble
at special viewing areas to see the colours of the rock intensify.

evening light – from the red end of the colour spectrum – on the rock and the surrounding sand.

Impressive though it is, I found that sharing it with the hundreds of camera and video-toting tourists that pack into the vast sunset viewing area each evening diminished the "wilderness" aspect of it. For a more intimate experience, try Kata Tjuta at sunset or, even better, Chambers Pillar or Rainbow Valley (see Chapter 11), or the Devils Marbles (Chapter 12).

Before leaving Uluru, be sure to visit Maruku Arts and Crafts, next to the ranger station, which features the work of 800 craftspeople living throughout the central and south-western deserts. The distinctive woodcarvings, decoratively etched with a hot wire, are displayed in a simple desert setting surrounded by traditional shelters, where some of the craftspeople give demonstrations.

Kata Tjuta (The Olgas)

The average visitor to Yulara spends only 1.8 days there, so many can only fit Uluru into their hectic schedules. That means those with the time to visit the awesome domes of Kata Tjuta, described by Giles as being "wonderful and grotesque", will often find few other people there.

Giles reckoned the many conglomerate domes, which rise up to 600 m from the plain, were "extraordinary freaks or convulsions of nature". He noted in his diary: "The appearance of this mountain is marvellous in the extreme, and almost baffles description … It is formed of several vast and solid rounded blocks of bare red-coloured stone, of a kind of conglomerate, being composed of untold masses of round stones of all kinds, mixed as plums in a pudding, and set in vast and rounded shapes upon the ground."

I found Kata Tjuta even more impressive than Uluru. The best way to explore the weathered domes is on the 8 km Valley of the Winds Walk, which circles some of the domes before leading up between some of them to the breathtaking Valley of the Winds Lookout. Apart from the short, steep section to the lookout, it's an easy walk that can be comfortably completed in three hours. Along the way I spotted a euro and her largish joey halfway up one of the domes.

If you begin the walk in mid-afternoon, you'll return in time to see the effect of the setting sun on Kata Tjuta. It's an experience you're likely to share with just a sprinkling of other people and consequently it's more enjoyable than its Uluru counterpart.

As you gaze at the setting sun, you'll be looking towards the Petermann and Mann ranges, where Harold Bell Lasseter claimed he'd discovered a rich gold reef in 1897. He died in 1931 while on an unsuccessful expedition to relocate it, taking the tantalising mystery of the reef to his grave.

Henbury Meteorite Craters Conservation Reserve

If you've ever wondered what the Moon's surface is like, a visit to the 12 meteorite craters at Henbury, 130 km south-west of Alice Springs, will give you a good idea. US

The complexity of the Kata Tjuta domes is even more impressive close up and an easy way to experience it is to follow the 2 km return track to Olga Gorge (above), or the 8 km Valley of the Winds Walk. Visitors with little time will get the best views of the boulders from afar towards sunset (opposite) and after storms (following pages). The thick clouds are the result of condensation as rain evaporates on the hot rock.

astronauts visited the site in the 1960s in preparation for the Moon landings, using the sparsely vegetated, pock-marked landscape to simulate the Moon's surface.

The craters were formed 4700 years ago by a meteor weighing several tonnes that plunged into the Earth's atmosphere at more than 40,000 km/h. It disintegrated before impact and the four main masses of metal – each about the size of a fuel drum – gouged out the largest crater, which is 180 m wide, 15 m deep and dotted with saltbushes. Smaller fragments formed other craters, some of which are only a few centimetres deep. More than 500 kg of metal has been found on the site, including one piece that weighed over 100 kg.

Jim's Place

The remarkable feature of this roadhouse on the Stuart Highway 89 km south of Alice is its owner, Jim Cotterill. His father Jack, who died in 1976, was one of the Red Centre's early pioneers of tourism and in 1960, Jack and Jim cut the road through to Kings Canyon and established Wallara Ranch at the junction of Ernest Giles and Luritja roads.

When Wallara was bulldozed in 1990 after Jim and his landlords failed to reach an agreement over the lease, he established Jim's Place, moving much of his tourism memorabilia – and his vast collection of photographs – with him. If you thought your trip to Uluru and Watarrka was an adventure, you'll think again after chatting to Jim.

Travel wise

The first 77 km of the Larapinta Drive are sealed, the remaining 36 km to Hermannsburg are unsealed but suitable for 2WDs. Access to Palm Valley and Boggy Hole is 4WD only and unsuitable for towing. Ernest Giles Road is unsealed east of Luritja Road but suitable for 2WDs, as is the access to Henbury meteorite craters. Towing is not recommended. The remainder of Ernest Giles Road and all of Luritja Road and the Lasseter Highway are sealed.

Entry to Uluru–Kata Tjuta National Park requires a $10 pass, valid for five days, obtainable from the park entry station.

Fuel, food and/or meals are available at Hermannsburg, Frontier Kings Canyon, Kings Creek station, Yulara, Curtin Springs Roadhouse, Mt Ebenezer Roadhouse, Desert Oaks Resort (Erldunda) and Jim's Place.

Accommodation and camping are available at Frontier Kings Canyon, Yulara, Curtin Springs, Mt Ebenezer, Desert Oaks Resort and Jim's Place.

There is camping at Palm Valley and Kings Creek station, basic camping (no water) at Henbury Meteorite Craters and bush camping at Boggy Hole.

Pedal power (below). English cyclist Tim Davies pedals away from Yulara on the Lasseter Highway, which was named after Harold Bell Lasseter, who claimed to have found a rich gold reef in the area in 1897. Possible locations for the reef include the dune country (opposite left) stretching west towards Bloods Ranges. On the way to Alice Springs from Yulara, Kings Canyon and South Australia, travellers pass a piece of memorabilia that's priceless to its owner, Jim Cotterill (opposite right). It's the old Dodge truck and steel drag Jim and his father Jack used to cut the first road into Kings Canyon in 1960 and it's one of many mementoes displayed at Jim's Place, a friendly roadhouse on the Stuart Highway 89 km south of Alice Springs. Equally entertaining are Jim's stories of early tourism.

4WD adventure in the Simpson Desert

Covering an area more than twice the size of Tasmania and stretching into Queensland and South Australia, the seemingly endless parallel ridges of the Simpson Desert are one of the remotest and least-visited features of the Red Centre. They're also the setting for one of the most remarkable tourist ventures in Australia – Old Andado, where Molly Clark (inset) dispenses old-fashioned outback hospitality to travellers who've followed the dusty 300 km 4WD

Molly Clark smiles when 4WD enthusiasts preparing to follow the 755 km Simpson Desert Loop ring her up to find out the condition of the first section of track from Alice Springs Airport to Old Andado, her tourist venture nestled in the Simpson Desert.

For Molly has been driving the 300 km track about once a month since she and her late husband Mac took over Andado cattle station in 1955, and she thinks nothing of it. "You carry food, water and a good book. If you run into problems you get in the back and read until someone comes along to help," she told me matter-of-factly as we tucked into lamb casserole in the homely kitchen of Old Andado. It's sound advice and, at 70, Molly is living proof that it works – she's had to follow it several times.

Glancing around the room as we chatted, I realised that everything – from the food we were eating and the delicate china it was served on to the fridge and freezer it had been stored in – had travelled down the rough track. Older items, like some of the multitude of clocks that each seemed to show a different time, would have been transported in the days when Molly had a two-wheel-drive vehicle – and well before she carried out track

improvements that lopped about two hours off the driving time. It was hardly surprising that she should be dismissive of what many travellers to the Red Centre regard as one of the great 4WD treks.

The Simpson Desert Loop, which veers southeast from Alice Springs Airport to Old Andado, south to Mount Dare, then north-west to Apatula, Chambers Pillar and back to Alice, is one of several 4WD adventures in the Northern Territory Tourist Commission's excellent booklet, *Tracks of Australia's Northern Territory*.

It takes travellers deep into the vast parallel sand ridges of the Simpson Desert through some of the most remote, timeless and deserted landscapes in Australia. On the day Tony and I drove to see Molly, we saw only a handful of vehicles on the first stretch to the Aboriginal community of Santa Teresa, and none on the final 234 km to Old Andado. And the next day we didn't see a single vehicle on the much more demanding 285 km of red sand track between Old Andado and Chambers Pillar.

More than anywhere else we visited in the Red Centre, this felt like frontier country – as wild and untamed as it was when the first European explorers traversed it in 1860.

track from Alice Springs. For $35 a night, she'll provide dinner, bed and breakfast, and as many anecdotes about her 40 years at the former cattle station as time permits. Judging by the testimonials in the visitors' book, it's a formula her guests love and many have become regulars, returning year after year for up to a week. "It's all word of mouth," she said. "I won't advertise or produce brochures because it's too darned expensive."

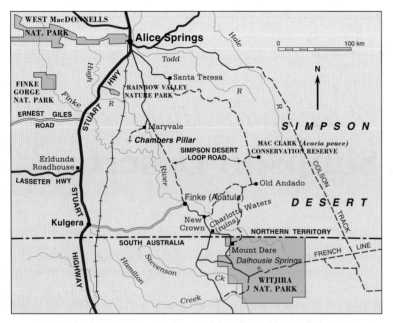

The Simpson Desert is best known for its vast sand ridges, some longer than 200 km, but adventurers will find plenty of variety as they follow the 755 km Simpson Desert Loop. The rough stone country (right) to the south-west of Old Andado is one such area of contrast and it highlights the harshness of the landscape and the need to be well prepared. Motorists should carefully calculate their fuel needs and plan their fuel stops before leaving Alice Springs and they should carry plenty of water – at least 10 litres a head for each day on the track – and food. In the event of a breakdown, they should stay with their vehicle, where they're more likely to be spotted, and keep as cool and calm as possible. It's a wise precaution to leave details of travel plans with a responsible adult who can raise the alarm if necessary.

The Old Andado Track

Molly Clark can drive between Alice Springs and Old Andado in four hours, but it's best to set aside a full day for a more leisurely journey with frequent stops to wander off the track and absorb the stark beauty of the ever-changing landscape.

The bitumen ends near Alice Springs Airport but the 66 km to the Allambi turn-off near Santa Teresa, a former Catholic mission, is on a well-graded track through open country dotted with rocky hills. The graceful ironwoods and wattles lining the track from Santa Teresa to Allambi station were flowering at the time of our journey and their cream and yellow blooms added splashes of bright colour to their dense foliage. Beneath them the red soil was littered with the sun-dried remains of everlastings, whose vivid yellow flowers had carpeted the landscape after the last good rains some eight months earlier.

Beyond Allambi the track deteriorates and there are several sections with deep bull-dust, but it improves again as it reaches the first dunes of the Simpson Desert, which, at 170,000 sq. km, covers an area more than twice the size of Tasmania. For most of the remainder of the journey to Old Andado the track follows a depression between two of the desert's 1100 or so parallel ridges, which run south-south-east into SA. At about 10 m high, they're small compared with others further east, which rise up to 40 m, but they're still impressive. Some are more than 200 km long.

Be sure to stop and clamber to the crests of some to gaze over the seemingly endless vistas of red sand. I was amazed by the number of delicate tracks criss-crossing the rippled sand between clumps of spinifex, parched grasses and tiny purple-flowered parakeelyas and sat quietly, hoping to see their creators – some of the 92 reptile and 44 mammal species living in the dunes. I saw none – apart from the ants on which many of them would feed – but there was something strangely comforting about seeing their tracks. The dunes may look inhospitable but they're home to a multitude of life.

The Mac Clark (Acacia peuce) Conservation Reserve

Some 250 km from the airport, the track enters desolate-looking gibber country as the dunes give way to vast ironstone plains. Tony thought it looked like the end of the world; to me it looked more like another planet. It was probably the most barren landscape we'd ever seen, yet we knew from our map that we were approaching the 20 km detour to the Mac Clark *(Acacia peuce)* Conservation Reserve, which protects a stand of rare waddy-wood trees.

From a distance, the spindly looking trees rise from the shimmering plain almost like a mirage. There are about 1000 of them, growing up to 17 m high in one of the hottest, driest parts of Australia where the only other vegetation we could see was the occasional stunted saltbush.

Tough as nails. The rare waddy-wood trees (above) that rise from the desolate gibber plains in the Mac Clark (ACACIA PEUCE) Conservation Reserve are tenacious survivors in what must be some of the toughest conditions in Australia. They take the extreme desert temperatures, frequent hot winds and 150 mm average annual rainfall in their stride, reaching heights of 17 m. Found in only two other places, near Boulia and Birdsville in south-west Queensland, they were felled by the early settlers for buildings and stockyards – until it was discovered they were too tough to drive a nail into. Generally, the gibber country (left) supports just a few stunted saltbushes and feels almost like the edge of the world, yet it's flanked by dunes that teem with plant and animal life.

Molly Clark's domain (above). Recent rains cover the Old Andado landscape with a mantle of green that seems as out of place in the desert as the rambling station buildings where Molly greets guests. More typical are the surrounding dunes (opposite above) that are home to some 92 reptile and 44 mammal species. They're rarely bare, but are usually dotted with spinifex and other grasses and, after rain, carpets of wild flowers. Molly has abundant water from an underground bore but uses it sparingly, tending her flower and vegetable garden with a watering-can.

Even though I found them less impressive up close, I couldn't help but admire these remarkable trees. Up to 500 years old, they survive on an average annual rainfall of just 150 mm and have small, needle-like leaves that minimise moisture loss through evaporation.

Aborigines carved clubs – known to the Europeans as waddies – from the dense timber. Pastoralist Mac Clark took an active interest in preserving the trees and when the 3042 ha reserve was declared in 1982, it was named after him.

Old Andado

I wasn't sure what to expect as I walked up to the faded green kitchen door of Old Andado, the rustic corrugated iron and coolibah-timbered homestead that's the hub of Molly Clark's low-key tourist venture nestled in a wide depression between two sand ridges.

I'd parked near a timber shed, thatched with brush, that was full to bursting point with the bits of machinery, vehicle spares and general clutter that outback folk keep because "they might come in handy one day". Nearby were rusted trucks and oil drums and, as I made my way to the front veranda, a jumble of old furniture. The only signs

that someone gave the place a little TLC were the patch of lawn and the neatly edged flower and vegie beds.

That someone is white-haired Molly, the proud and sprightly owner of Old Andado, who greeted me as I stepped into the welcoming kitchen. I saw at once that she channelled her considerable energy into maintaining the interior of the 1922 property – and greeting a steady stream of visitors – rather than beautifying the exterior, which she prefers to keep looking much as it would have done three generations ago.

I paused near the doorway. An immaculate iron cooking range dominated the far end of the room, quaint dressers lined the corrugated iron wall opposite me and an embroidered tablecloth and dainty china were spread on the table that filled the middle of the room. The aroma of fresh baking was everywhere.

When Molly sold Andado in 1985, she kept the best bit – Old Andado, which she had moved to with Mac in 1955. They lived in it with their three sons until they built a bigger homestead in 1960, and from 1972 Molly opened it to tourists to show them how the outback pioneers lived. In 1985 she moved back as live-in curator.

Her guests stay in one of five twin-bedded rooms at the back of the five-room homestead, or in the camp ground to the side. "We get as many overseas tourists as we

The five-roomed corrugated iron and timber homestead (below) that's the hub of Old Andado was built in 1922. Molly moved in with her late husband Mac in 1955, shortly after the dirt floor of the homely kitchen (opposite) had been covered with cement and electricity had been installed. The room now has a modern cooker, fridge and freezer at the opposite end to the old-fashioned range (inset), on which Molly does most of her cooking – except in summer. Otherwise little has changed – and that's the way Molly and her guests like it. "It reminds me of visits to my grandmother when I was a little girl," one elderly guest commented. Molly stocks up on food in Alice Springs every four or six weeks, usually spending at least two nights away so she can catch up with friends, so it's a good idea for intending visitors to ring her to make a firm booking.

do Australians," she told me as we relaxed after dinner in big armchairs in front of the fireplace in the cosy lounge. "I don't count them – I've got more to do than count tourists – but a lot come back, even overseas people. We've had some German people three times, which I think is pretty good."

As I looked around the room with its full bookshelves, old photographs and a lifetime's mementoes, I realised that the attractions of Old Andado were threefold: its history, its isolation and, last but definitely not least, its owner. Molly's a true Aussie battler who's had more than her fair share of personal tragedy and hardship, yet she refuses to give in. Although Old Andado doesn't make a profit, she's vowed to keep going as long as she can.

"People think you're getting rich because you have a few tourists going through, but they've got no idea what it costs to live here," she said, reeling off some examples: $120 in petrol every time she drives to Alice Springs for supplies, $400 for two drums of fuel to run the generator for a month and $1000 to get someone in to fix that generator.

When Molly's on her own she sometimes uses just kerosene lamps, but the generator's generally on for four hours each evening and all of every other night in summer to keep the freezers going. She has a phone and a radio, but no television or newspapers – and clearly doesn't miss them.

When rain falls in the Simpson Desert, it can be heavy enough
to form large pools that will gradually leach into the sand. For
creatures like burrowing frogs, it means a rare opportunity to emerge
from their water-holding shroud beneath the sand to feed and breed.
And for many seeds laying dormant below the surface, it means a
chance to germinate and bloom. From the air, the parallel ridges of
the desert are much more striking. They run north-north-west from
South Australia into the Northern Territory and were formed by
the prevailing winds.

DICK SMITH

DICK SMITH

"People often ask me if I feel frightened being out here all on my own," she said. "But what have I got to be scared of here? There's plenty in town, but not out here."

In the tracks of The Ghan

From Old Andado, the Simpson Desert Loop goes west to Andado, then south to Mount Dare before heading north-west to New Crown station. But as this book's boundary is the South Australian border, we skipped the Mount Dare section and travelled south-west over the 15 m crests of the wind-rippled dunes to New Crown, where we spotted several camels resting in the shade of coolibahs near a dry swamp. The track often appears daunting, but it's been improved in recent years and the 120 km from Old Andado to Apatula – the latter part through flatter, wooded country punctuated with mesas – is generally good.

Permits aren't needed to visit Apatula, once an important supply point for the old Ghan train, and it's a good place to top up with fuel and food, and stretch your legs before the long, slow drive to Chambers Pillar. Aboriginal artefacts made by some of the settlement's 190 residents are well worth looking at.

As soon as the track leaves Apatula, the real 4WD adventure begins when you're faced with the wide, soft, sandy bed of the Finke. If you've never driven in sand before, think seriously before undertaking the remaining 165 km through 5–8 m red sand ridges to Chambers Pillar. Even if you have experience, you'll be grinding along at 20 km/h for most of the journey and however intense your concentration, you're bound to get bogged occasionally (see *Getting around*, Chapter 13).

It's at this point that the track meets the old narrow-gauge Ghan railway line, which opened in 1929 to link Alice Springs with the railhead at Oodnadatta. It closed in 1980, when the new standard-gauge line opened, and although few of the rails are left, most of the heavy timber sleepers remain. As I inspected some just beyond the Finke, I recalled a conversation I'd had the previous week with Laurie Nicholls, who worked on the Ghan on its weekly run for four years from 1947 and now spends his holidays taking tourists along a short stretch of restored track near Alice Springs.

He'd told me about one trip when the driver drove the engine onto a timber bridge crossing the Finke – and found himself in more than a metre of water. "We got pulled out into a siding and had to wait 11 days for the water to go down," he said, adding philosophically: "We were lucky. If we'd got there half an hour later there would have been 20 ft (6 m) of water." Gazing at the parched landscape around me, his story would have been hard to believe if I'd not already witnessed the power of the Finke some 225 km to the north-west at Finke Gorge.

At Bundooma Siding, 85 km from Apatula, I stopped near a stand of desert oaks for a closer look at a huge overhead water tank, built to supply the Ghan, and the ruins of an old settlement. A little further on, soon after veering left at a Y junction 1.5 km beyond the siding, I spotted the first of several poles that once were part of the Overland Telegraph Line. I couldn't help but admire the tenacity of the pioneers who'd

Between Apatula and the Chambers Pillar turn-off, the soft sand of the Simpson Desert Loop becomes something of a 4WD challenge and each June, over the Queen's Birthday weekend, part of it is used by trail-bikers in the Finke Desert Race. This aerial view (opposite) shows the Finke River in the background and the old Ghan and desert loop tracks in the foreground. Timber sleepers (below) and a few sections of track are all that remain of the old Ghan line, which closed in 1980. The track to Chambers Pillar is busier and fairly firm (right), but still negotiable by 4WD only.

TONY STANTON

travelled this route on horseback in 1872 to establish this link between Australia and the rest of the world.

Chambers Pillar

The sight of Chambers Pillar and other weathered sandstone formations like Castle Rock rising from the vast red plain that stretches to the horizon is one of the most haunting in the Red Centre. I arrived shortly before sunset, in time to sit on a low sand ridge dotted with desert oaks and watch the sun's dying rays light up the whites, yellows and reds of the 50 m high column with an almost fiery intensity.

The pillar was named in 1860 by John McDouall Stuart, the first European to see

it. He described it simply as "a remarkable hill" and likened it to "a locomotive engine with its funnel". Ernest Giles was more eloquent. On 23 August 1872 he noted: "The appearance of this feature, I should imagine, to be unique in Australia, and it is not likely that any future explorer will ever discover so singular a monument, wherewith to immortalise either himself or his patron."

I sat gazing at the pillar until well after dark, imagining I was seeing it exactly as Stuart and Giles had done more than 100 years earlier. We were travellers of different eras: they were on horseback; I was in an airconditioned 4WD. They relied on compass bearings; I had maps and a global positioning system. But I liked to think that despite our differences, we shared a similar sense of adventure and regarded the pillar with similar awe.

A strong sense of history washed over me again the following morning when, shortly after sunrise, I clambered up the steep hill from which the pillar rises for a closer look at its base. Many of the early explorers carved their names into the sandstone and among the first initials I spotted were those of pioneer pastoralist William Hayes. Only a few days earlier I'd been chatting to his great-grandson Jim in the kitchen of Undoolya homestead on the outskirts of Alice Springs. Another great-grandson, Billy, runs Maryvale station, which I'd passed on the drive to the pillar.

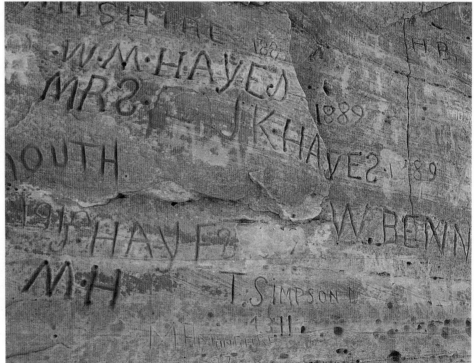

Many modern-day adventurers want to leave their mark on the pillar too, but carving more names undermines the historical significance of those already there, so the practice has been declared illegal and several offenders prosecuted. If you want to record your visit, sign the visitors' book in the camping area a five-minute walk from the pillar.

You could visit Chambers Pillar on a day trip from Alice Springs, but I'd strongly recommend an overnight stay in the delightful camping area to see it come alive at sunset and sunrise, and to enjoy the sense of solitude. It was with a sense of regret that, like Giles, "we turned our back upon this peculiar monument and left it in its loneliness and its grandeur".

Ewaninga Rock Carvings Conservation Reserve

The circles and animal tracks etched into soft sandstone outcrops beside a claypan in this 6 ha reserve are believed to be some of the oldest rock carvings in the Red Centre, possibly dating back 20,000 years.

Also clearly visible in the reserve, which is reached some 125 km after leaving Chambers Pillar to follow the Simpson Desert Loop back to Alice Springs, are grooves in the rock where stone axes were ground.

The glorious vista (below) of Castle Rock, on the left, and Chambers Pillar rising from the plain is ample reward for negotiating the rough terrain to reach them. Both sandstone formations have resisted erosion because they're protected by hard capping, and they're at their most impressive at first and last light (opposite left). Chambers Pillar is probably the most historic landmark in the region, having acted as a beacon to scores of explorers, surveyors and settlers since being discovered by John McDouall Stuart in 1860. Many, including several members of the Hayes pastoral family, carved their names in the soft rock and they read like a WHO'S WHO of Red Centre pioneers. Today's explorers are asked to leave their mark in a visitors' book so they don't detract from the historical significance of those on the rock.

PIP AND DICK SMITH

Rainbow Valley

Alice Springs locals regard Rainbow Valley, a majestic sandstone bluff that forms an impressive backdrop to a vast claypan, as one of the Red Centre's best-kept secrets. I confess I found it a little disappointing, but that was because I made the mistake of arriving in the early afternoon of a particularly hot day. The spinifex-dominated landscape offered no shade and it was too hot to tackle any of the walks to Aboriginal engravings and paintings nearby, so I kept my visit brief.

Later I learnt that I should have stayed to see the late afternoon sun light up the rainbow-like bands in the rock, which were formed by the action of water. The red iron in the sandstone dissolved in wetter times and during dry seasons was drawn to the surface, where it hardened. It weathers much more slowly than the softer white sandstone beneath it. Had I stayed, perhaps I would have considered the rock as impressive as Chambers Pillar, as many locals do.

The bluff's admirers say it's particularly impressive after rain, when its reflection can be seen in the water trapped in the claypan. Wild flowers like yellow tops, brought on by the rain, add to the spectacle.

Rainbow Valley can't be reached by the Simpson Desert Loop. The only access is

via an unmarked track to the east of the Stuart Highway 75 km south of Alice Springs. As you head south it's just past a cattle grid, but if you miss it, carry on to Jim's Place, a roadhouse 14 km south of the turn-off, and get directions there.

Travel wise

The Simpson Desert Loop is unsealed. The section between Apatula and Chambers Pillar is 4WD only and unsuitable for towing. A 4WD is recommended for the remainder of the loop and towing is not advisable, although it is possible between Alice Springs and Old Andado. A 4WD is recommended for the last 22 km to Rainbow Valley.

A permit is necessary to visit Santa Teresa and is available from the Central Land Council.

Accommodation and meals are obtainable at Old Andado by prior arrangement with Molly Clark on (089) 56 0812. Camping is also available.

Fuel is sold at New Crown, Apatula and Maryvale; food is available at Apatula.

There are basic camping facilities (no water) at Chambers Pillar and Rainbow Valley.

The usually deserted claypan in front of the striking cliffs of Rainbow Valley is transformed into a hive of activity (above) in preparation for Corroboree 94, a travel convention organised by the Australian and Northern Territory tourist commissions. The claypan, which visitors are now asked to avoid walking on, looks very different when heavy rain turns it (opposite above) into a shimmering lake, dominated by a huge whitewood. For adventurers looking for an alternative way to see the area to the south-east of Alice Springs, a camel trek (opposite below) is a challenging option. Frontier Camel Farm offers a five-and-a-half-day trek that takes riders into the Simpson Desert and to Chambers Pillar, enabling them to sleep out under the stars and see the landscape much as some of the early explorers did.

Gems and other treasures on the road north

Balancing act. The Devils Marbles are probably the most bizarre of the Red Centre's many remarkable rock formations and it's not surprising that some visitors think they're too strange to be natural. The hundreds of weathered granite rocks are up to 6 m in diameter and are scattered haphazardly over a wide area – often in precarious-looking heaps. Originally huge

"The good news is that you're going to find a lot of zircons today," Graham Short said confidently as we rummaged through a pile of wet and glistening pea-sized stones. I'd just plucked out my first zircon, a glassy, semi-precious stone that resembles a diamond, and visions of glittering necklaces were flashing through my mind.

Graham quickly brought me down to earth. "The bad news is that very few will be of gem quality – maybe only one or two out of 20 or more," he said. But I wasn't too concerned. I was only five minutes into a gem fossicking lesson with him, I had a full day ahead of me, and Tony to share the hard, physical work. If I found just one zircon I could have cut and set into a piece of jewellery, I'd be more than happy.

We were 92 km north-east of Alice Springs at Mud Tank zircon field, a barren-looking dust bowl where the mineral-rich ground has yielded some of the world's largest zircons.

Our lesson began with some tips on finding a suitable site. Using a pickaxe, Graham chipped at the red-brown earth, looking for black stones called ironstone. "Where we find ironstone, we should find zircons because they're the same weight," he explained.

When he had found a suitable patch of easily visible ironstone, he chipped to a depth of 15–20 cm, shovelled the dirt into a large sieve and shook it vigorously to get rid of the fine dust. He then transferred the contents into a smaller sieve and dunked it in water to wash the stones, which he tipped onto a board for us to examine.

"This is where we need sunlight, which we haven't got today," he said, pushing back his battered bush hat and eyeing the clouds. "We're looking for something glassy and if the sun's shining, it picks it up. It's harder without it."

Having satisfied himself that we knew what we were looking for, Graham left us and for the next five hours we chipped, shovelled, shook, washed and scrutinised batches of dirt. It was hard, hot, dirty work, but thoroughly absorbing. Several times successive sievefuls yielded nothing and we'd be on the verge of packing up when we'd find something promising in what we'd told ourselves would be the last sieveful. The urge to try just one more lot was overwhelming.

We eventually amassed about 40 stones that we thought were zircons and drove the 15 km to Graham's base at Gemtree Caravan Park for his expert assessment.

chunks, they have gradually been rounded through erosion by wind and rain. The "marble" on the left of this pile will gradually become two smaller marbles as water penetrates and erodes the large vertical fracture through its middle. But it will be a lengthy process – the erosion visitors see the results of today has been taking place over the past 310 million years.

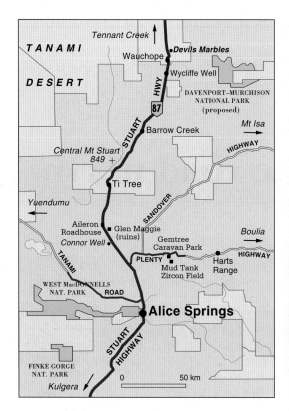

Glittering prizes (right). Graham Short, in the centre, shows eager gem fossickers how to spot a glassy, semi-precious zircon as he sorts through washed stones at Mud Tank zircon field. Gem fossicking is hard, dirty work (above right), but it's strangely addictive and Graham virtually guarantees that everyone will find at least one zircon that – when cut, polished and set into jewellery – will sparkle like a diamond. Zircons are among the many gemstones found in the mineral-rich Harts Range to the north-east of Alice Springs.

Gemtree

Fossicking enthusiasts Graham and June Short established this caravan park, general store, fossicking centre and gem shop in 100 ha of bush beside the Plenty Highway in 1989 as a base for the growing number of fossickers visiting the Harts Range area to search for zircons, garnets, quartz, beryl and a host of other semi-precious stones.

Since then they've built up a thriving business giving lessons at $30 a time, which includes equipment hire, and cutting and polishing stones – and they've introduced hundreds of people to the hobby they stumbled upon by chance when they gave up their careers in the computer industry 10 years ago to travel around Australia.

"One day in Queensland, someone suggested we look for garnets near Cloncurry," said Graham. "We thought it was fun and we came out here looking for zircons and enjoyed that too. We spent the next six months going to different gem spots around Australia." He learnt gem cutting and spent four years working on the gem fields in the Harts Range as a cutter before establishing Gemtree.

"We have people who stumble across us for a couple of days because someone told them about us," he said. "The next year they may come back for two weeks and the following year for three or four weeks. Other people come here for weeks and weeks at a time."

I could understand why. There's something strangely addictive about fossicking because you never quite know what you're going to find. I watched eagerly as Graham held each of our stones to a powerful light and examined them through magnifying glasses. "I'm looking for clearness," he explained. "A lot are cracked and flawed and won't look good or may fall apart when they're cut."

He selected 10 as being of gem quality and although none looked particularly impressive in the raw state, I knew the true beauty would be revealed when they were cut and polished. I left three with George Weier, Gemtree's resident cutter, to have set into a pendant and ring.

George, who describes his work as "a hobby that got out of control", takes about an hour to machine-cut an average zircon into the 65 or 73 facets that make the stones sparkle and he charges $14 a carat, the unit by which gemstones are weighed. He's worked at the park for four years and says the biggest zircon he's cut was 21.4 carats – about 2 cm long. Mine, at between 0.6 and 1.8 carats, were small by comparison and their value, about $70 a carat for a cut stone, was modest. But I reckon the finished jewellery looks a million dollars.

The road north

Most travellers heading north from Alice Springs on the Stuart Highway have one objective: to get to the Top End of the Territory. In winter, when overnight temperatures in the Red Centre frequently drop to below freezing, most want to reach the tropical north as quickly as possible.

That was my aim when I first ventured north of Alice in 1991. Apart from regular

At his base at Gemtree Caravan Park (above), Graham Short scrutinises the washed zircons for clarity. Clear, uncracked stones can be machine-cut by George Weier, in the background, so they sparkle from 65 or 73 facets. At the height of the season Graham employs two other cutters to cope with the demand from fossickers eager to convert their finds into attractive pieces of jewellery as souvenirs of their visit to the caravan park.

Roadside stalls selling rockmelons, pumpkins and other fruit and vegetables 180 km north of Alice Springs are the most visible signs of a remarkable farming enterprise in the very centre of Australia. It was started in 1975 by Ti Tree farmer Ian Dahlenburg (above), who pioneered the cultivation of table grapes on former mulga scrubland and was joined three years later by his son Arthur and daughter-in-law Diane. They've since added asparagus and figs to their repertoire and now employ up to 10 people to harvest asparagus twice a year to send to the southern capitals. Farming in one of the driest parts of Australia has not been without its problems – one year, after a freak hailstorm, Ian received just $30 for his grape harvest. "All the trellis was damaged and our melons and pumpkins were bruised too," Ian said. "We had to give away pumpkins for cattle feed."

stops to change drivers, Tony and I drove the 393 km to Devils Marbles ignoring the various points of interest as the incredibly straight road traversed the mostly flat, mulga-dominated scrub. Later I learnt we'd missed some interesting snippets of history and ignored two fascinating business ventures. Apart from Gemtree, which is a 70 km detour from the highway, you pass within a few metres of them all and although most merit little more than a brief stop, they're a good opportunity to stretch your legs.

Like most roads in the Red Centre, this stretch of the Stuart Highway is never busy and it's not hard to picture the scene in 1907, when the first motor vehicle to attempt to cross Australia from south to north came face to face with a cyclist doing the journey in reverse. The car, driven by geologist Henry Dutton and his mechanic Murray Aunger, who succeeded in their endeavour the following year, encountered the bicycle, ridden by adventurer Francis Birtles, near Barrow Creek and the trio shared a billy of tea before going their separate ways.

The "road" wasn't even a ribbon of dirt in those days, but in the late 1930s and early 1940s a single lane was constructed in preparation for the hundreds of military convoys that headed north from the railhead, troop reserve and arsenal established in Alice Springs to service the war effort further north.

From 1940, almost 200,000 troops travelled up the highway, which was sealed in 1942. Connors Well and Ti Tree, where water was available, became lunch stops, Barrow Creek was developed as an overnight staging camp, and the army established a farm at Wycliffe Well. Wauchope also played a key role in the war effort: wolframite mined there was used in the manufacture of armaments.

Ryan Well

This lonely outpost beside the Stuart Highway 120 km north of Alice was one of at least a dozen wells in the Red Centre sunk by a government team, led by Irishman Ned Ryan, in the five years from 1885–90.

Work started in August 1889 and after four months of digging the 1.8 by 1.2 m shaft with picks and shovels and hauling out the dirt with a rope and bucket, the team struck water. The first travellers to use it hauled water up the 24 m timber-lined shaft using a hand windlass and bucket, but a horse whip and 45,000-litre tank were soon added to make drawing water simpler. Later, the owners of Glen Maggie station, which was opposite the well, charged passing travellers a fee to draw water for their stock.

In the 1930s, the spread of motorised transport and machine-drilled bores robbed the well of its earlier importance, but the well, the stone supports for the storage tank and the timber supports for the stock-watering trough still remain as memorials to Ned Ryan, who devoted nearly 30 years of his life to opening up the outback for European settlement.

Be sure to cross the road for a closer look at Glen Maggie homestead. The stone walls of the two front rooms are all that remain of the isolated sheep and cattle station established here in 1914 by Sam Nicker and his family. From 1921 the property served as a telegraph office and store and from 1932 it was the last supply point for miners seeking gold at The Granites, 800 km to the north-west. It was abandoned in 1935.

Ti Tree

"All this was just mulga scrub when I came here in 1975," Ian Dahlenburg told me as we walked through his 16 ha asparagus plantation at Ti Tree Farm, 180 km north of Alice Springs. Slender asparagus spears were sprouting among the delicate fernlike leaves that arched over the red sandy loam and, in the distance, pickers were snapping off the 20 cm long shoots.

"We pick twice a year – from late March to early May and from the end of July to September," he said, offering me a tender spear to munch as we walked. "At the beginning of the season, it's often picked twice a day because it can grow more than 5 cm a day."

I was witnessing a flourishing farming operation in one of the remotest and most unlikely parts of Australia – less than 40 km from the spot John McDouall Stuart calculated to be the centre of the continent and in an area with an average annual rainfall of just 250 mm.

Ian, lean and wiry with a weather-beaten face from a lifetime on the land, pioneered fruit and vegie growing in the Red Centre in 1975 when he planted table grapes on part of a 260 ha plot. Despite early setbacks caused by nematodes (parasitic worms), watering problems and a freak hailstorm that left him with just $30 worth of grapes one season, he persevered and he now has 5 ha of grapes and 4 ha of recently planted figs, as well as asparagus.

The Red Centre's desert climate – abundant sunshine and cool winter nights – means the crops mature at different times to those in other parts of Australia, so the growers are able to recover the high costs of getting their produce to markets in Adelaide, Melbourne, Sydney and Perth by selling it at a premium. Ian's grapes begin ripening in November, well before those further south, and his success has lured other farmers to the area, including a major grape-growing concern whose 100 ha site nearby produces a crop worth about $3 million a year.

I left Ian to grade the freshly picked asparagus that would net him up to $50 for a 6 kg box and drove down the track to see his neighbour, John Crayford at Red Centre Farm.

John, a burly ex-plumber, bought his 10 ha plot from Ian six years ago and specialises in mangoes, a fruit he admits he'd never tasted when he planted the first of his 1000 trees. As he led me along the neat rows of lush, glossy-leaved trees, he told me his 1994 crop yielded 600 trays of fruit, which he sold for $18–$20 a tray.

"My market is Adelaide," he said. "If they want good mangoes they have to pay a high price for them because they are so far away from other growers. That puts me in a good position, despite the high labour and freight costs."

John plans to concentrate solely on mangoes in the future, but in the meantime he's

Plumber turned farmer John Crayford (above) checks the flowers on one of his 1000 mango trees, which flourish in the red sandy loam near Ti Tree thanks to abundant bore water. "The only problem with growing mangoes here is that some of the fruit gets sunburnt because it's on the tree at the hottest time of year," he said. The fruit is hand-picked in January and February when it's still green, graded in John's packing room and stored in a cool room until it's trucked to Adelaide. "Our mangoes are sweeter than any in Australia," John said, adding that some weigh up to 700 grams. Like many of his neighbours, John receives help and encouragement from the Department of Primary Industries, which has a nearby arid zone research station where experimental plantings of citrus fruits and native flowers are being evaluated.

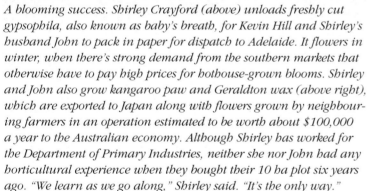

A blooming success. Shirley Crayford (above) unloads freshly cut gypsophila, also known as baby's breath, for Kevin Hill and Shirley's husband John to pack in paper for dispatch to Adelaide. It flowers in winter, when there's strong demand from the southern markets that otherwise have to pay high prices for hothouse-grown blooms. Shirley and John also grow kangaroo paw and Geraldton wax (above right), which are exported to Japan along with flowers grown by neighbouring farmers in an operation estimated to be worth about $100,000 a year to the Australian economy. Although Shirley has worked for the Department of Primary Industries, neither she nor John had any horticultural experience when they bought their 10 ha plot six years ago. "We learn as we go along," Shirley said. "It's the only way."

growing flowers, many of which are exported to Japan, as a cash crop. "We've got a lot of kangaroo paw – mainly yellows but a few reds and a black variety. They start flowering towards the end of September and I send a few to Adelaide but most go to Japan at $1 a stem."

It was hard to imagine such beautiful flowers blooming in what was once scrub in one of the driest parts of Australia, but John told me: "It's good farming land. We don't rely on rain – there's [bore] water – and you can grow anything here."

Central Mount Stuart

John McDouall Stuart was disappointed when he encountered this dome-shaped rocky hill in 1860. It was some 4 km south-south-west from his calculated centre of Australia and he noted: "I wish it had been in the centre."

On 23 April 1860 he recorded in his diary: "Took [William] Kekwick and the flag, and went to the top of the mount, but found it to be much higher and more difficult of ascent than I anticipated. After a deal of labour, slips, and knocks, we at last arrived on the top."

He and Kekwick built a large cairn on the summit and in its centre placed a pole with a British flag nailed to it. "We then gave three hearty cheers for the flag, the emblem of civil and religious liberty, and may it be a sign to the natives that the dawn of liberty, civilisation, and Christianity is about to break upon them," he wrote.

He named the peak Mount Sturt, but it was later renamed in Stuart's honour. It's on

pastoral property and a memorial on the Stuart Highway is the nearest most travellers get to it.

Barrow Creek

The most substantial building in this tiny settlement is the telegraph station that was built in 1872 as a repeater station on the Overland Telegraph Line. Its stone walls and corrugated iron roof were designed to resist attacks by Aborigines, but in 1874 officer-in-charge James Stapleton and linesman John Franks were killed in an ambush that led to scores of Aborigines being shot in reprisal.

With its wide verandas and backdrop of tall mesas, the building is well worth closer inspection. As well as being used as a post and telegraph office and for housing station staff and linesmen who worked on the line, it later served as an Aboriginal ration depot, a police station and a staging camp for army convoys during World War II, when it also played an important role in providing military communications. Inadequate water supplies led to the staging camp being moved to a site 30 km further north in 1942.

Devils Marbles Conservation Reserve

I couldn't help but feel humbled as I walked among the bizarre, brilliant orange-red boulders that comprise the Devils Marbles. Some of the weathered granite rocks were more than three times my height and several had two or three other boulders perched

The Creek Bank (above) at the Barrow Creek roadhouse rewards "investors" in a novel way. Patrons are invited to sign a banknote and pin it on the bar wall so that if they ever return thirsty and broke, they can still enjoy a drink. The landscape around the tiny settlement, which also includes an old telegraph repeater station, is characterised by expansive mesas (above left).

*The Devils Marbles in perspective. A ghost gum (above) and an
enthusiastic visitor (opposite) show the scale of the giant boulders
that are among nature's most impressive creations.*

precariously on top of them. Spindly ghost gums sprouted from cracks above my head and tiny pigmy goannas darted in and out of crevices lower down. This was nature on a grand scale.

None of the dozens of photographs I'd seen of the Devils Marbles – most of them focusing on one or two particularly dramatic formations – had prepared me for the sheer scale of them. Up to 6 m in diameter, they were much bigger than I expected and were scattered for several kilometres across a wide, shallow valley where the only vegetation was golden spinifex and various wattles.

Towards dusk I spent more than an hour wandering among the marbles, which were originally much bigger granite tors that have been rounded and reduced by erosion and the gradual flaking away of thin layers from their surface. The setting sun intensified their rich colours, which are the result of staining by irons and other minerals.

Be sure to allow an hour or two to wander among these rocks, which Aborigines say are the eggs of the rainbow serpent and, if possible, time your visit for shortly before sunrise or sunset, when they undoubtedly look their best.

Davenport/Murchison National Park

This rugged 1100 sq. km national park now being developed is renowned for its permanent waterholes, which have long been popular with locals. Due to open in 1996, the park will be low-key to retain the wilderness atmosphere and will cater mainly for experienced bushwalkers.

Road trains are a familiar sight on the Stuart Highway and many other Red Centre roads and play a vital role in bringing in food and supplies and transporting cattle. Motorists should pass these giants of the road slowly and give them as wide a berth as possible, particularly on dirt roads. David Parkinson (above), one of seven drivers working for Tanami Tranport, is often away from his Alice Springs home for a week or more as he transports cattle from stations around Alice Springs to Port Augusta in the south or, more commonly, to Darwin in the north for shipment to Asia. His 50 m road train can carry 150 grown cattle or up to 300 weaners and he's come to know the Stuart Highway as well as anyone. Like most travellers, he appreciates landmarks like the Devils Marbles (opposite and above left), which add interest to the long journey north.

Travel wise

The Stuart Highway is sealed, as is the Plenty Highway as far as Gemtree. The track to Mud Tank zircon field is unsealed but suitable for 2WDs.

You don't need a Miner's Right to fossick at Mud Tank, but you need the $20 licence – valid for a lifetime and obtainable from the Department of Mines and Energy in Alice Springs (089) 52 3658 – to fossick on the many nearby sites in the Harts Range that are crown land.

Fuel, food and/or meals and accommodation and camping are available at Aileron Roadhouse, Ti Tree, Barrow Creek, Wycliffe Well, Wauchope and Gemtree.

There are basic camping facilities (no water) at Devils Marbles.

HOWARD WHELAN

Scenery like the magnificent Chewings Range in the West MacDonnells National Park awaits visitors to the Red Centre. With so much to see and do in the region, it's a good idea to research and plan your itinerary carefully before leaving home.

Planning your visit

The more carefully you research your Red Centre adventure, the more rewarding it will be. There's nothing worse than discovering when you're back home that you missed a beautiful waterhole, a fascinating tour or an experience you would have cherished forever, so it's well worth taking the time and effort to find out about the region and what it has to offer before you leave home.

The most common mistake visitors make is trying to cram too much into too little time – and spending most of their visit rushing from one place to another without seeing anything properly. Perhaps more than anywhere else in Australia, you need time to savour the unique atmosphere of the Red Centre, to seek out those remote wilderness spots and to discover some of the elusive wildlife.

To visit most of the places described in this book and have time to do some of the walks, I'd recommend a minimum of three weeks. A month would be even better. If you can only spare a week or a fortnight, concentrate on exploring a few places thoroughly and save the rest for another visit.

Getting there

Unless you're lucky enough to live in the Red Centre or on its fringes, or you decide to fly there, just getting there can be quite an expedition. The road journey from Perth to Alice Springs is 3933 km; from Sydney it's 2958 km, from Brisbane 2846 km, from Melbourne 2259 km, from Canberra 2680 km, from Darwin 1490 km and from Adelaide 1530 km.

If time is no object, consider following one of the "adventure" routes – like the Plenty Highway or the Strzelecki, Birdsville and Oodnadatta tracks – for part of the way, or breaking your journey at Coober Pedy or one of the many other places of interest you'll pass. Another alternative is to travel from Adelaide to Alice Springs on the Ghan train, which also has space for cars.

Getting around

The big advantage of driving yourself is that you have a car when you reach the Centre – and you can pack swags, camping gear and as many creature comforts as space allows.

If you travel by plane or train but want to explore independently when you arrive, cars, 4WDs and camper vans can be hired at Alice Springs and Yulara. Alternatively there are a host of organised tours ranging from adventurous camping safaris to luxury coach trips with comfortable motel accommodation.

Road improvements over the past decade have made most of the Red Centre's attractions accessible by 2WD and it's even possible to tow a caravan or trailer to many. However, a 4WD is essential to visit Finke Gorge National Park, Ruby Gap Nature Park, Rainbow Valley, Chambers Pillar and for the Boggy Hole Track and Simpson Desert Loop. They can be hired in Alice Springs, but drivers should have some experience of 4WDs – or have been on a specialist driving course – because they may see few other vehicles.

Driving in sand can be tricky and it's a good idea to carry sand mats, available from 4WD accessory shops. If you find yourself getting bogged, slip into reverse gear and, if possible, try backing out, then engage first gear and edge slowly forward. Sand mats or a layer of twigs and leaves under the tyres will help improve traction, as will lowering tyre pressures.

Travelling safely

The sense of isolation that makes the Red Centre so special for many visitors means you need to be well prepared, whether you're exploring it on foot or by vehicle.

If you're walking, carry – and drink – up to a litre of water for every hour on the track, and be well equipped with protective clothing, a first-aid kit, maps, compass and spare food. Rangers ask all walkers on the Larapinta Trail to register with them and even on shorter walks it's a good idea to leave details of your route and intended return time with a responsible adult. Don't forget to notify them of your safe return.

If you're motoring, calculate your fuel, water and food needs carefully before setting off – allowing extra fuel for heavy loads, roof-racks, towing, headwinds and dirt roads, and at least 10 litres of water a day for each person. Bear in mind that, except where clearly indicated otherwise, distances given in this book are "as the crow flies" and that road distances may be up to 50 per cent longer. Before venturing off the beaten track, check conditions with police, rangers and other travellers and, again, leave details of your travel plans and intended return time with a responsible adult.

Always carry maps, two spare tyres and a puncture repair kit, pump, valve keys and pressure gauge, spark plugs, fanbelt, jumper leads, radiator hoses, tow rope, insulating tape, water and fuel cans and a funnel, fire extinguisher, spade and axe. If you plan to spend a lot of time in remote spots, consider hiring a high-frequency radio for contact with the RFDS.

If you break down, keep calm, ration food and water *and stay with your vehicle*, where you'll be much easier to find.

Caring for the bush

Many visitors to the Red Centre like to imagine they're seeing the landscape exactly as the first European explorers did 130 years ago, but the illusion is easily shattered if they encounter rubbish. The message is simple: if you can carry it in, you can carry it out, however you're travelling. Don't be tempted to bury rubbish – animals will soon dig it up and scatter it around, and may harm themselves in the process.

Much desert wildlife relies on fallen trees and branches for shelter, so firewood should be gathered sparingly, and not at all from national parks and nature reserves, where its collection is prohibited. If you enjoy a campfire on a chilly evening, buy firewood from a garage, keep your fire small and never leave it unattended because most of the vegetation is highly flammable.

If pit toilets are available, use them. If not, dig toilet holes well away from watercourses and waterholes and carry out toilet paper, condoms, disposable nappies and sanitary products.

Where to stay

The Red Centre is among the best places in the world to unroll your swag and sleep under the stars, listening to the bush come alive with the howls of dingoes. Rain is unusual and provided you have warm clothing and sleeping-bags for chilly nights, the camping is second to none.

For those who prefer a roof over their heads, Alice Springs has a range of accommodation, from caravan parks and backpackers' hostels to self-catering apartments and motels. Yulara has a similar range, but elsewhere the options are more limited. Advance booking is recommended for Yulara and Kings Canyon because much of their accommodation is block-booked by tour companies.

When to go

The cooler season from April to September is the most comfortable. Average maximum temperatures range from about 20°C in June and July up to about 30°C in April and September, providing ideal conditions for walking and exploring.

Dirt roads are common in the Red Centre but many can be safely negotiated by 2WD.

The desert climate means winter nights can be chilly – often below freezing – so if you're camping, take plenty of warm clothes and good-quality sleeping-bags.

Temperatures during the summer months often top 40°C and are too hot for safe and comfortable walking between 10 a.m. and 4 p.m. However, if you don't intend to do much walking and will be travelling and staying in airconditioned comfort, March and October are worth considering. Rainfall is most common from December to March.

What to wear

The fierce Red Centre sun demands respect throughout the year, so a wide-brimmed hat, sunglasses and sunscreen are essential. Unless you intend visiting Lasseters Casino or dining in one of the best hotel restaurants, comfortable, casual clothing is all that is needed. Shorts and a T-shirt, or thin cotton trousers and long-sleeved shirts that offer protection from the sun, are ideal day wear in all but the coldest months, when trousers and a sweater are often necessary.

Take warm clothes for the evening and early morning, even if you're not camping. They're essential for dinners under the stars and early-morning walks in near-freezing temperatures. Rainwear is rarely needed.

Strong shoes or walking boots are recommended for the rocky terrain, so too are gaiters if you plan to do a lot of walking. A fly veil offers welcome relief when the bush flies get you down.

Photography

There's nothing like a set of stunning photographs to keep your holiday memories alive and the Red Centre is a photographer's dream.

Barry Skipsey, who took the magnificent photographs for this book, suggests you always keep your camera at the ready, because you never know what you'll encounter.

The optimum times for landscape photography are early morning and late afternoon, ideally the first and last 10 minutes of light. "You can't beat being there when the light is perfect, however sophisticated your camera and equipment are," Barry said. Between those times a polarising filter will eliminate glare and make the colours brighter and richer. If you're photographing people, particularly Aborigines, use fill-in flash to highlight facial features and a polarising filter to reduce shine on skin surfaces.

Barry uses a 28 mm lens for about half of his work and finds it ideal for showing people and their environment. Other favourites are a 135 mm for portraits and a 350 mm and 500 mm for wildlife and for picking out details from the landscape. He finds a slow-speed film works best in the strong sun and favours Fujichrome Velvia, which is 50 ASA. To compensate for the slow speed he recommends using a sturdy tripod and cable release wherever possible.

Three final tips from Barry, who's spent 14 of his 18 years in the Territory photographing the Red Centre: never change films in a dusty environment because grit can get into the camera and cause scratches, keep camera gear and film out of direct sunlight and heat as much as possible, and always pack more film than you think you'll need.

Aboriginal land

Much of the land in the Red Centre is Aboriginal land and permits are necessary. Transit permits are required to use some roads that cross Aboriginal land (including the road to Tnorala and the Mereenie Loop Road) and entry permits are essential for visiting many Aboriginal communities. Details and application forms are available from the Central Land Council, PO Box 3321, Alice Springs, NT 0871. Ph: (089) 52 6320.

Many Aboriginal communities have been declared "dry" areas and visitors should not take in alcohol or consume it there. They should also seek permission before taking photographs on Aboriginal land.

Banks

Alice Springs has Commonwealth, Westpac and National bank branches. Electronic banking facilities are available at Yulara and Frontier Kings Canyon. Credit cards are widely accepted.

General

Prices of goods in the outback tend to be high because of the cost of transporting them. At the time of writing, unleaded petrol was 63c a litre in Sydney; 83c in Alice Springs and up to $1 elsewhere.

Further information

The tourist information produced by the Northern Territory Tourist Commission is among the best in Australia. So too are the national parks leaflets published by the Conservation Commission of the Northern Territory and the Australian Nature Conservation Agency, which manages Uluru–Kata Tjuta National Park. A charge is made for some leaflets, but most are free.

Once you've arrived in the Centre, the Central Australian Tourism Industry Association office on the corner of Hartley Street and Gregory Terrace, Alice Springs, is the best place for information on tours. It also has a CCNT counter.

Northern Territory Tourist Commission
67 Stuart Highway
Alice Springs
NT 0870
Ph: (089) 51 8555

Conservation Commission of the Northern Territory
PO Box 1046
Alice Springs
NT 0871
Ph: (089) 51 8211

Australian Nature Conservation Agency
PO Box 119
Yulara
NT 0872
Ph: (089) 56 2299

Central Australian Tourist Information Association
PO Box 2227
Alice Springs
NT 0871
Ph: (089) 52 5199

Further reading

The most useful books to take with you as you explore the Red Centre are the *Wildlife Identikit* and *Plant Identikit* published by the CCNT for $4 each. They are small enough to fit in a shirt pocket and cover the most common wildlife and plants, although more detailed guides are handy to keep in your car. I also enjoyed consulting the published diaries of John McDouall Stuart and Ernest Giles, which I bought in Alice Springs for $35 each.

Index

Main references are in **bold type.**
Italic indicates a caption, but references occurring in both text and caption on the same page are in roman type only. An asterisk indicates another feature at the locality referred to, e.g. a walking track or historic site.

Abbott, John 107
Aborigines 15, 59, 62, 79, 80, *93*, 100, *103*, 104, *115*, 116, 123, 128, 152, **165**, 169, 173
art *33, 34, 35*, 105, 107, 117, 127, 137, 155, 156
history **25–37**
implement making **132–4**
outstations 35–7
Adelaide House 76, 77
Afghans *46, 86, 103*
Alice Springs 42, 45–47, **75–89**
Alice Springs Desert Park 78
Amburla Brahman Stud 55
Anderson, Keith 77
Andrews, James 44
Appel, Kristen 111, 112
Arenge Bluff *93, 97*
Arltunga Historical Reserve 45*, 46*, **111–3**, 115*
Aunger, Murray 162
Auricht, Glen 35, *36, 37*
Australian Geographic *19, 126*, 127, 128

Bagot, Edward *44*
Barrow Creek Telegraph Station 30, 162, **165***
Battarbee, Rex *105*
Birtles, Francis 76, 162
Boggy Hole 31, 119, 120, 123
Bond Springs 86, 89
British-Australian Telegraph Company 39, 42
Brooks, Frederick 31
Brown, Chris "Brownie" *19*

Bundooma Siding 152
Burke and Wills 40
Bush tucker *25, 37, 93, 97, 107*
Byrnes, Steve 57

Camel catching 55, *126*
Camel Cup *86*, 89
Camel trekking *88, 109, 157*
Carmichael, Samuel 42
Castle Rock 153, *155*
Cattle industry 45, *51*, 52, *55*
Central Australian Aviation Museum 76, 77
Central Land Council 33, 173
Central Mount Stuart 40, **164–5**
Chambers Pillar 23, 39, *41*, 42, 137, 143, 152, **153–5**, 156, *157*
Chateau Hornsby 83–4, *85*
Chewings Range 68, 91, *94, 170*
Chisholm, Georgina 55
Chisholm, Sam *55*
Churchill-Smith, James *44*, 45
Clark, Mac 143, 146, *148*
Clark, Molly 143–4, **146–8**, 152
Cleland Hills 25
Climate 67, *111, 119, 145*, 161, 163, 172
Community Development Employment Program *36, 37*, 107
Coniston massacre 31, 84
Coniston station 31
Connellan, Eddie 47, 76
Connors Well 162
Conservation Commission of the Northern Territory 56, 97, 173
Conway, Ian and Lyn 52, 55, *126*, 127
Corroboree Rock *115*, 116
Cotterill, Jack 47, 49, 140
Cotterill, Jim 47, 49, 140, *141*
Cox, Philip 128
Crayford, John 163–4
Crayford, Shirley *164*

Dahlenburg, Ian *162*, 163
Dann, Tanya *55*
Dann, Troy *55*, 55
Darwin Overland Maintenance Force 46
Davenport Murchison National Park 169
Deep Well station 45
Defence 52, **57**
Department of Primary Industries *163, 164*
Desert Tracks 37
Desert Wildlife Park and Botanic Gardens 56
Development **51–8**
Devils Marbles 15, *20*, 23, 137, *159*, 162, **165–9***
Dunes 23, 71–2, 123, *143*, 144–5, *150*
Dutton, Henry 162

Easter 1988 floods *103, 120*, 123
Egan, Ted 84, *85*, 86
Ellery Creek 17, 18, 123
Ellery Creek Big Hole 18, 100, 109
Emily Gap 116
Ernest Giles Highway 127
European history **39–59**
Ewaninga Rock Carvings Conservation Reserve 155

Finke Desert Race *153*
Finke Gorge National Park 42, **120**, 123,
Finke River 18, *23, 41*, 44, 67, 100, *103, 106*, 120, 123, 152, *153*
Finlayson, Sister 76
Flint, Ernest *30*
Flynn, John 76, *77, 102*
Fox's grave 115
France, Ted 51–2
Franks, John *30*, 165
Frontier Camel Farm *157*
Frontier Kings Canyon 127

Garden of Eden 125
Garnet 115
Gemtree Caravan Park 159, **161**, 162
Geology **15–23**
George Gill Range 44, 125, 127
Ghan Preservation Society 83
Ghan, the *75*, 171 (*see also* Old Ghan)
Ghost Gum Lookout *100*
Gibber *73*, 145
Gibson Desert 45
Gibson, Alfred 44, 45
Giles, Ernest 29, *39*, **42–5**, 119, *120*, 123, 125, 134, 137, 154
Glen Annie Gorge 115
Glen Helen Gorge 18, 100, *103, 106*
Glen Helen Homestead 47, 100, *102, 103, 108, 109*
Glen Maggie station 162
Glen of Palms 44, 123
Gold 45, **111–3**, 115
Gosse, William 44–5
Great Western Mine 111, *112*, 113
Guth, Henk 80

Hale River 113, 115
Harts Range *160*, 161
Hayes, Billy 154
Hayes, Gail *44*, 45, 52
Hayes, Jim 44, 45, 52, 154
Hayes, William *44*, 45, 154, *155*
Head, Benjamin 39
Heaslip, Brett *53*
Heaslip, Grant 86
Heaslip, Jan 86, 89
Heavitree Range 91, *94, 97, 98*
Heenan, Brendan 75
Heenan, Mick 75
Henbury Meteorite Conservation Reserve **137, 140**
Henley-on-Todd Regatta 89
Hermannsburg 31, *79*, 80, **104–5**
Hermannsburg Potters *106*

Hitchcock, Bobby *77*
Hope, Andrea *93*
Horn Scientific Expedition *124*
Hornsby, Denis 83–4, *85*
Horticulture 55, **162–4**
Hot-air ballooning *88*, 89
Howard, Chris *93, 97*
Hugh River 18
Hugo, David 79

Illamurta Springs 31, 123
Inarlanga Pass 18, 91, *92, 93, 97*
Introduced animals 59, *64*, **65**, 116, 119–20, *123*

James Range 107
Jay Creek 93
Jessie Gap 17, 116
Jim's Place **140**, *141*, 157
Jindalee Over-the-horizon Radar Network 57
John Flynn 77
John Hayes Rockhole 116, *117*
Johnston, George 47
Joint Defence Facility Pine Gap 57

Kalarranga Lookout *23*, 120
Kantju waterhole 134
Kata Tjuta 15, 17, 18, 42, **137***
Kekwick, William 39, 40, *41*, 164
Kempe, Hermann 31
Kidman, Sidney 86
King, Ian 125, 127
Kings Canyon 23, 42, 44, 49, 104, 123*, *124*, 125*, 126, 140, *141*
Kings Creek station 52, 55, *126*, 127
Kookaburra 76, 77
Kurkara Tours 127

Lake Amadeus 44
Langford, Mike 128
Larapinta Drive 104
Larapinta Trail 91, *92*, 93, 100

Index

Lasseter Highway 127, 128, *141*
Lasseter, Harold Bell *80*, 137, *141*
Lasseters Hotel Casino *80*
Levi Range 125
Liddle, Arthur 49
Lindsay, David 115
Liru Walk 132–4
Lost City 125
Luce, Henry 113
Luritja Road 127

Mac Clark (*Acacia peuce*)
    Conservation Reserve **145–6**
MacDonnell Range Reef Mine
    111, 113
MacDonnell Ranges 15, 17, 18,
    19, 39, 42, 44, 45, 47, 91, 97,
    104, *111*, 115
MacDonnell Ranges cycad 68, 91
Mala Walk 134
Maruku Arts and Crafts 35, 137
Maryvale station 45, 154
Matthews, Dennis *123*
McFadzean, Gary *56*
Mecca Date Farm *88*
Mereenie Loop Road 104
Mesa 23, *41*
Moon landings 140
Mpaara Track 120
Mpulungkinya Track 120
Mt Connor *126*, 128
Mt Ertiva Siding 80, *85*
Mt Giles *94*, 100, *101*
Mt Olga 44, 45
Mt Razorback 91
Mt Sonder *19*, *94*, *101*
Mt Sturt 164–5
Mt Undoolya *111*
Mt Zeil 15, *17*
Mud Tank zircon field 159, *160*
Mueller, Johannes 75
Museum of Central Australia 89
Museum of Technology,
    Transport and
    Communications 76, 79

Musgrave Ranges *25*

N'Dhala Gorge Nature Park
    116, **117**
Namarari, Mick *34*
Namatjira Drive 100, 104
Namatjira, Albert *91*, *105*
Napperby station *55*
Natjuna, Maureen 132–4
Nautiloid *19*
New Crown station 152
Nicholls, Laurie 80, 83, 152
Nicker, Sam 162
Northern Territory Tourist
    Commission *157*, 173

Oil and gas production 52, **56**
Old Andado 143, 144, **146–8,**
    152
Old Ghan, the 80, 83, 152–3
Olga Gorge *157*
Olive Pink Flora Reserve 89
Ormiston Gorge *94*, 100, *101*,
    102
Orr, Allen 83
Overland Telegraph Line 29, *39*,
    42, 45, 152, 165

Palm Valley 23, 47, *105*, 120*
Palmer River 123
Panorama Guth 80
Papunya Tula *34*, 35
Parkinson, David *169*
Passmore, Gary 93, 97
Petermann Ranges 18, *80*
Photography tips 172–3
Pioneer Women's Hall of Fame 56
Pitchi Richi Sanctuary 89
Plenty Highway 161
Porter, Glenys 107
Porter, Ken 107
Puritjarra shelter 25
Railways 45–6 (*see also* Ghan)
Rainbow Valley 23, 137, **156–7**
Rainfall 15, *103*, 146, 163

Red Centre Farm 163–4
Redbank Gorge 100, *102, 109*
Responsible camping 172
Road trains 79, *169*
Robinson, Alexander 42
Rocky Creek 97
Ross Highway **115–6**
Ross River 47
Ross River Homestead 18, 116
Ross, John 42, 115
Royal Flying Doctor Service **57,**
    89
Rubies 115
Ruby Gap Nature Park **113, 115**
Running Waters station 123
Ryan Well **162–3**
Ryan, Ned 162

Safe travel 171
Salt lakes 73
Sand plains 23, 71–2
School of the Air **51–2,** 89
Schwarz, Wilhelm 31
Serpentine Chalet 47, 91, *92*, 100
Serpentine Gorge 68
Short, Graham 159, *160*, 161
Short, June 161
Shute, Neville 75
Simpson Desert 15, *115*, 116, 120,
    143, 144, 145
Simpson Desert Loop **143–57**
Simpsons Gap *89*, 91, 100*, *102*
Skipsey, Barry *13*
Smith, Dick *11*, 76
Smith, Mike 25
Spencer, Baldwin *26*, 29
Standley Chasm 104, **108**
Stanton, Jenny 11, *13*
Stapleton, James *30*, 165
Stokes Creek 127
Strehlow Research Centre 79–80
Strehlow, Carl *79*, 80
Strehlow, Ted 79, 80
Stromatolites 18
Stuart (Alice Springs) 45, 46, 83*

Stuart Highway 42, *46*, 47, *75*,
    161*, 162, *169*
Stuart Town Gaol 76
Stuart, John McDouall 26, 29,
    39–42, 153–4, *155*, 163, 164
Sturt, Charles 39

Tanami Transport *169*
Tangentyere Council 33
Travel permits 173
Tempe Downs station 125
The Creek Bank *165*
Ti Tree 162, 163
Ti Tree Farm 163
Tietkens, William 44
Tjuwanpa Resource Centre 35,
    36
Tnorala (Gosse Bluff) 15, **100,**
    104
Todd River 75, *88*, 89
Todd, Charles 42
Tourism **47–9,** *51*, 52, **55–6,** *75*,
    86, 116, 125, 128, 140, *141*
Tracks of Australia's Northern
    Territory 143
Trephina Gorge Nature Park
    115, **116*,** *117**
Tyler Pass 100

Uluru (Ayers Rock) 15, 17, 18,
    44, 45, 47, 49, 119*, 128,
    132–5, 137
Uluru Experience 134
Undoolya station *44*, 45, 52, 154
Ungwanaka, Rachel *106*

Vegetation 15, *23*, 25, 26, *57*,
    67–73, *75*, 91, *92*, 97, *98*, 107,
    112, 115, 116, *117*, 119, 120,
    123, 125, *126*, 127, 133, 144,
    145, 152, 156, 162, *166*, 169
Wallace Rockhole 104, **107**
Wallara Ranch 49, 140
Watarrka National Park *43*, 49,
    119, **123–7***

Wauchope 162
Weier, George 161
West MacDonnells National
    Park 68, **91–103,** *170*
Wheelchair access 127
Whelan, Howard 127
Whillock, Kit 33, 127
White Range Mine 113
Wildlife 15, *57*, **59–65,** 72, 97,
    100, *101*, 115, 116, *117*, 119,
    120, *123*, 125, 145, *150*, 169
Williams, Barry 57
Williams, Jack 76
Willshire, William 119
Wine production 83–4
Winnecke *46*, 111, 112
World War II 33, **46–7,** 162, 165
Wycliffe Well 162

Yowa Bore 49
Yulara *26*, 49, **127–8***

Zircon 159, *160*

Makeshift road signs on each side of a dangerous bend in the Mereenie Loop Road, south-west of Alice Springs, display their warning message with a typical Territorian directness.